PENGUIN BOOKS

OTHER WORLDS

Paul Davies is Professor of Mathematical Physics at the University of Adelaide. He obtained a Ph.D. from the University of London and has held academic appointments at the universities of London, Cambridge and Newcastle-upon-Tyne. He emigrated to Australia in 1990. His research interests are in the field of black holes, cosmology and quantum gravity, and he has published over one hundred specialist papers as well as several textbooks.

Professor Davies has achieved an international reputation for his ability to explain the significance of advanced scientific ideas in simple language. He is the author of some twenty book, including the widely acclaimed *Superforce*, *The Cosmic Blueprint*, *The Runaway Universe* (Penguin 1980), *God and the New Physics* (Penguin 1984), *Other Worlds* (Penguin 1988), *The Mind of God* (Penguin 1993) which was shortlisted for the 1993 Science Book Prize, and, with John Gribbin, *The Matter Myth* (Penguin 1992). *The Edge of Infinity* is forthcoming in Penguin.

He is well known for his media appearances in several countries and has written and presented a number of TV and radio programmes, including a major series of documentaries on BBC Radio 3. In 1991 he was awarded the Eureka Prize for the promotion of science in Australia.

Paul Davies

OTHER WORLDS

PENGUIN BOOKS

PENGUIN BOOKS

Published by the Penguin Group
Penguin Books Ltd, 27 Wrights Lane, London W8 5TZ, England
Penguin Books USA Inc., 375 Hudson Street, New York, New York 10014, USA
Penguin Books Australia Ltd, Ringwood, Victoria, Australia
Penguin Books Canada Ltd, 10 Alcorn Avenue, Toronto, Ontario, Canada M4V 3B2
Penguin Books (NZ) Ltd, 182–190 Wairau Road, Auckland 10, New Zealand

Penguin Books Ltd, Registered Offices: Harmondsworth, Middlesex, England

First published in Great Britain by J. M. Dent & Sons 1980
Published with a Postscipt in Pelican Books 1988
Reprinted in Penguin Books 1990
10 9 8 7 6 5

Printed in England by Clays Ltd, St Ives plc
Set in VIP Bembo

Contents

Illustrations

Preface

Although the word 'quantum' has now entered popular vocabulary, few people are aware of the revolution that has taken place in science and philosophy since the inception of the quantum theory of matter at the beginning of the century. The stunning success of this theory in explaining molecular, atomic, nuclear and subatomic particle processes often obscures the fact that the theory itself is based on principles which are so astonishing that their full implications are often not appreciated, even by many professional scientists.

In this book I have tried to face squarely the full impact of fundamental quantum theory on our conception of the world. The behaviour of subatomic matter is so alien to our common-sense perspective of nature that a description of quantum phenomena reads like something from *Alice in Wonderland*. The purpose of the present book, however, is not simply to review a notoriously difficult branch of modern physics, but to turn instead to broader issues. What is man? What is the nature of reality? Is the universe we inhabit a random accident or the outcome of a delicate selection process?

The question of why the cosmos has the particular structure and organization that we observe has long intrigued theologians. In recent years, discoveries in fundamental physics and cosmology have opened up the prospect of a scientific approach to some of these questions. Quantum theory has taught us that the world is a game of chance, and we are among the players; that other universes could have been selected, and may even exist in parallel with our own, or in remote regions of spacetime.

The reader need have no previous knowledge of science or philosophy. Although many of the topics treated here require some mental gymnastics, I have tried to explain each new detail from scratch in the most elementary language. If some of the ideas seem hard to believe, it is testimony to the profound changes in the scientific view of the world that have accompanied the great progress of the last few decades.

By way of acknowledgment I would like to say that I have enjoyed fruitful discussions with Dr N. D. Birrell, Dr L. H. Ford, Dr W. G. Unruh and Professor J. A. Wheeler on much of the subject-matter treated here.

Prologue – the unnoticed revolution

Scientific revolutions tend to be associated with a major restructuring of human perspectives. Copernicus' claim that the Earth did not occupy the centre of the universe began a disintegration of religious dogma that tore Europe apart; Darwin's theory of evolution upset centuries of belief in the special biological status of humans; Hubble's discovery that the Milky Way is but one among billions of galaxies scattered throughout an expanding universe opened up new vistas of celestial immensity. It is therefore remarkable that the greatest scientific revolution of all time has gone largely unnoticed by the general public, not because its implications are uninteresting, but because they are so shattering as to be almost beyond belief – even to the scientific revolutionaries themselves.

The revolution concerned took place between 1900 and 1930, but over forty years later controversy still rages over precisely what it is that has been discovered. Known broadly as the quantum theory it began with an attempt to explain certain technical aspects of subatomic physics. Since then it has grown to incorporate most of modern microphysics, from elementary particles to lasers, and nobody seriously doubts that the theory is correct. What is at issue are the extraordinary consequences that follow if the theory is taken at face value. If accepted completely literally, it leads to the conclusion that the world of our experience – the universe that we actually perceive – is not the only universe. Co-existing alongside it are countless billions of others, some almost identical to ours, others wildly different, inhabited by myriads of near carbon-copies of ourselves in a gigantic, multifoliate reality of parallel worlds.

To avoid this startling spectre of cosmic schizophrenia, the theory can be interpreted more subtly, though its consequences are no less mind-boggling. It has been argued that the other universes are not real, but only contenders for reality – failed alternative worlds. However, they cannot be ignored for it is central to the quantum theory, and can be checked experimentally, that the alternative worlds are not always completely disconnected from our own: they overlap our perceived universe and jostle its atoms. Whether they are only ghost worlds, or as real and concrete as our own, our universe is actually only an infinitesimal slice from a gigantic stack of cosmic images – a 'superspace'. The coming chapters will explain what this superspace is, how it works, and where we, the inhabitants of superspace, fit in.

Science, it is usually believed, helps us to build a picture of objective reality – the world 'out there'. With the advent of the quantum theory, that very reality appears to have crumbled, to be replaced by something so revolutionary and bizarre that its consequences have not yet been properly faced. As we shall see, one can either accept the multiple reality of the parallel worlds, or deny that a real world exists at all, independently of our perception of it. Laboratory experiments performed in the last few years have demonstrated that atoms and subatomic particles, which people usually envisage as microscopic *things*, are not really things at all, in the sense of having a well-defined, independent existence and a separate, personal identity. Yet we are all made of atoms: the world about us seems to be directed inevitably to an identity crisis.

In the mid 1960s a remarkable mathematical formula was discovered by the physicist John Bell. Any logical theory based on the independent reality of sub-atomic particles which adheres to the well-established principle that faster-than-light signalling is impossible obeys this formula. Quantum theory on the other hand does not. Recent experiments in which pairs of photons (particles of light) are sent simultaneously through two pieces of polarized material set obliquely to one another confirm that Bell's formula is indeed violated.

These studies show that reality, inasmuch as it has any meaning at all, is not a property of the external world on its own but is intimately bound up with our perception of the world – our presence as conscious observers. Perhaps more than anything else this conclusion carries the greatest significance of the quantum revolution, for unlike all the previous scientific revolutions, which have successively demoted mankind from the centre of creation to the role of mere spectator of the cosmic drama, quantum theory reinstates the observer at the centre of the stage. Indeed some prominent scientists have even gone so far as to claim that quantum theory has solved the riddle of the mind and its relation to the material world, asserting that the entry of information into the consciousness of the observer is the fundamental step in the establishment of reality. Taken to its extreme, this idea implies that the universe only achieves a concrete existence as a result of this perception – it is created by its own inhabitants!

Whether these latter paradoxical ideas are accepted or not, most physicists seem to agree that, at least on the atomic level, matter remains in a state of suspended animation of unreality until an actual measurement or observation is performed. We shall examine in detail this curious state of limbo, which corresponds to atoms being caught between many worlds, undecided where to go. We shall ask whether the limbo state is confined to the subatomic domain, or whether it can erupt out into the laboratory and percolate the cosmos. The famous paradoxes of Schrödinger's cat, and Wigner's friend, in which a man is apparently put in a 'live-dead' state and asked to relate his sensations, will be explored in an attempt to pin down the true nature of reality.

Central to the quantum theory is the inherent uncertainty of the subatomic world. The desire to believe in determinism, where every event has cause in some earlier event and the world unfolds in a tidy, law-abiding scheme, is deep rooted, and forms the basis of many religions. Albert Einstein clung closely to this belief all his life, and could not accept the quantum theory in its conventional form, for the quantum revolution injects an element of chance into nature at its most fundamental level. We all know that life is a chancy business and that we can never hope to predict accurately the future of complicated systems such as the weather or the economy, yet most laymen believe the world is in principle predictable, if only enough information were available. Physicists used to believe that even atoms obeyed the rules and moved about according to some precise system of activity. Two centuries ago, Pierre Laplace remarked that given knowledge of every atomic motion, the entire future of the universe could be mapped out.

The discoveries that took place in the first quarter of this century revealed that there is a rebellious aspect to nature. Intruding into what seems to be a law-abiding cosmos is a randomness – a sort of microscopic anarchy – that destroys the clockwork predictability and imbues the world of the atom with an absolute uncertainty. Only the laws of probability regulate an otherwise chaotic microcosm. In spite of Einstein's protestation that God does not play dice, it appears that the universe is a game of chance, and that we are not just spectators but players. Whether it is God, or man, who tosses the dice, turns out to depend on whether multiple universes really exist or not.

Chance or choice – is the universe we actually experience an accident, or have we selected it from a bewildering array of alternatives? Surely science can have no more urgent a task than to discover whether the structure of the world about us – the arrangement of matter and energy, the laws they obey, the quantities that have been created – is just a random quirk, or a deeply meaningful organization of which we form an essential part. In the later sections of the book some radical new thoughts on this subject will be presented in the light of the most recent discoveries in astrophysics and cosmology. It will be argued that many of the features of the universe which we observe cannot be separated from the fact that we are alive to observe them, for life is very delicately balanced in the scales of chance. If the multiple-universes idea is accepted, we as observers have *selected* a remote and tiny corner of superspace which is completely uncharacteristic of the rest, an island of life among chasms of uninhabited dimensions. This raises the philosophical problem of why nature has so much redundancy built in to it. Why produce so many universes when all but a tiny fraction go unnoticed? If, instead, the other universes are relegated to ghost worlds, we must regard our existence as a miracle of such improbability that it is scarcely credible. Life is then indeed chancy – more chancy than we could ever conceive.

The inherent uncertainty of nature is not confined to matter, but even controls the structure of space and time. It will be shown that these entities are not merely the stage on which the cosmic drama is acted out, but belong to the cast. Space and time can change their shape and extension – crudely speaking, they move about – and like subatomic matter their motion is somewhat random and uncontrolled. We shall see how on an ultra-microscopic scale the uncontrolled motions could tear space and time apart, endowing it with a sort of frothy, foam-like structure, full of 'wormholes' and 'bridges'.

Our experience of time lies closest to our perception of reality, and any attempt to build a 'real world' must come to grips with the paradoxes of time. The most profound puzzle of all is the fact that, whatever we may experience mentally, time does not pass, nor does there exist a past, present and future. These statements are so stunning that most scientists lead a sort of dual life, accepting them in the laboratory, but rejecting them without thought in daily life. Yet the notion of a moving time makes virtually no sense even in daily affairs, in spite of the fact that it dominates our language, thoughts and actions. It is here, perhaps, that new developments lie, in unravelling the mystery of the linkage between time, mind and matter.

Many of the topics treated in this book are 'stranger than fiction', yet the remarkable thing is not their peculiarity, but the fact that the scientific community has known of them for so long without making much attempt to communicate them to the public. Probably the chief reason is the exceptionally abstract nature of the quantum theory, plus the fact that it is usually approached only with the help of very advanced mathematics. Certainly a lot of the topics in the coming chapters will test the imagination of the reader, but the issues are so profound and important to us all, that some attempt at bridging the gulf of understanding must be made.

1. *God does not play dice*

In the early 1920s an American physicist, Clinton Joseph Davisson, began a series of investigations for the Bell Telephone Company in which nickel crystals were bombarded by a beam of electrons, similar to the beam which produces an image on a television screen. He noticed some curious patterns in the way the electrons scattered off the crystal surface, but did not then understand their enormous significance. Several years later, in 1927, Davisson conducted an improved version of the experiment with a junior colleague, Lester Halbert Germer. The pattern was very pronounced, but more important was the fact that it was now expected, on the basis of a remarkable new theory of matter developed in the mid-1920s. Davisson and Germer were directly observing for the first time a phenomenon that brought about the collapse of centuries of entrenched scientific belief, and turned upside down our notions of the meaning of reality, the nature of matter and our observation of it. Indeed, so profound is the revolution in knowledge which followed and so bizarre are the consequences, that even Albert Einstein, perhaps the most brilliant scientist of all time, refused throughout his life to accept some of them.

The new theory has now become known as quantum mechanics, and we shall explore the astonishing implications it has for the nature of the universe and our own role in it. Quantum mechanics is not merely a speculative theory of the subatomic world, but an elaborate mathematical framework which holds together most of modern physics. Without quantum mechanics our detailed and extensive understanding of atoms, nuclei, molecules, crystals, light, electricity, subatomic particles, lasers, transistors and much else would disintegrate. No scientists seriously doubt that the basic ideas of quantum mechanics are correct. However, the philosophical implications of the theory are so startling that even after fifty years, controversy still rages over what it really means. To appreciate the profundity of the quantum revolution, it is first necessary to understand the classical picture of nature as conceived by scientists at least since the seventeenth century.

In the earliest days, when men and women first began to wonder about the natural events that went on around them, their picture of the world was quite different from ours today. They realized that some events were regular and dependable, like the days and seasons, the phases of the moon and the motions of the stars, while others were arbitrary and apparently random, such as storms, earthquakes and volcanic eruptions. How were they to organize this knowledge into an explanation of nature? In some cases, a natural happening would have had an obvious explanation; for example, when the warmth of the sun melted snow. But a precise notion of cause and effect was not well formulated. Instead, it must have seemed very natural to model the world on the system that they understood best – themselves. It is easy to see how natural phenomena came to be regarded as a manifestation of *temperament* rather than causality. Thus, regular and dependable events reflected a placid commitment and benign method at work, while sudden and perhaps violent events were seen as the product of a petulant, angry or neurotic temperament. One outcome was astrology, in which the order apparent in the heavens was taken as a reflection of a wider organization, linking celestial and human temperament in a unified scheme.

In some societies the temperamental systems crystallized somewhat, and became actual personalities. There was the wood spirit, the river spirit, the fire spirit, and so on. The more developed societies constructed an elaborate and strongly anthropomorphic hierarchy of gods. The sun, moon, planets – even the Earth itself – were regarded as humanlike personalities, and the events which befell them a reflection of well-known human emotions and desires. 'The gods are angry' would have been regarded as sufficient explanation for some natural calamity, and appropriate sacrifices made. The power of these grander personalities was taken very seriously, probably amounting to the greatest single sociological force.

Paralleling these developments was a new collection of ideas induced by the development of city dwelling and the appearance of nation states. To prevent anarchy citizens were expected to conform to a strict code of conduct which became institutionalized into *laws*. As expected, the gods were also subject to laws and in turn, by virtue of their greater power and authority, endorsed the human system of laws with the help of their intermediaries, the priests. In the early Greek civilization the concept of a lawful universe was very far advanced. Indeed, the explanations for routine natural occurrences, such as the flight of a missile or the fall of a stone, were starting to be formulated as unfailing *laws of nature*. This dazzling new concept of phenomena that acted unsupervised strictly in accordance with natural law stood in sharp contrast to the alternative idea of an organic world regulated by motivated temperament. Of course, the really important phenomena – the astronomical cycles, the creation of the world and man himself – still required the close attention of the gods, but the commonplace could look after itself. However, once the idea of a physical system evolving automatically according to a fixed and inviolable set of principles took root, it was inevitable that the domain of the gods should be progressively eroded as more and more new principles came to be discovered.

Although the retreat of the theological explanation for the physical world is even today not complete, the decisive steps in establishing the power of physical laws came, broadly speaking, with Isaac Newton and Charles Darwin. During the sixteenth century, the intellectual giant, Galileo Galilei, began what today we would call a series of laboratory experiments. The key idea was that by isolating a piece of the world as much as possible from surrounding influences, it would be free to behave in a very simple way. This belief in simplicity at the heart of complexity has been a driving force behind scientific inquiry for millenia, and persists undiminished today, in spite of the shocks that, as we shall see, it has received in recent times.

One celebrated investigation that Galileo conducted was to observe the progress of falling bodies. Usually this is a very complex process depending on the weight, shape, mass distribution and internal motion of the body as well as the wind speed, density of the air, and so on. Galileo's genius was to spot that all these features are only incidental complications superimposed on what is really a very simple law. By reducing the effect of air resistance and using bodies of regular shape, by rolling them down inclines (rather than dropping them directly) so as to simulate the effect of greatly reduced gravity, Galileo managed to cut through the complexity and isolate the fundamental law of falling bodies. What he did in essence was to measure the time required for bodies to fall different distances. This may seem a reasonable enough procedure today, but in the seventeenth century it was a masterstroke of genius. The concept of time in those days was quite different from our own: for example, the notion of a mathematically regular passage of time was not accepted. Temporal duration had long been closer to the old organic ideas, deriving concreteness more from the natural rhythms of the human body, the seasons, or the heavenly cycles, than from precision clockwork. With the discovery of America, and the establishment of routine transatlantic passages, strong commercial and military pressure stimulated the search for more accurate east-west navigation procedures. It soon became apparent that by a combination of accurate star-fixing and time measurement, a ship's longitude in mid-ocean could be computed. Thus began the construction of observatories and the science of modern positional astronomy, as well as the invention of ever-more accurate clocks.

Although it was still more than a generation before Newton formalized the concept of an 'absolute, true and mathematical time' and all of two centuries away from the railway schedules which finally brought this concept into the lives of ordinary folk, Galileo correctly identified its central role in a description of the phenomenon of motion. His reward was the discovery of a disarmingly simple law: the time to fall a given distance from rest is exactly proportional to the square root of that distance. Science was born. The idea of a *mathematical formula* rather than a god supervising the behaviour of a physical system had arrived.

The impact of this development cannot be overstated. A law of nature as a mathematical equation not only implies simplicity and universality, but also tractability. It meant that it was no longer necessary to observe the world to ascertain how it would behave: you could also compute it with a pencil and paper. Using mathematics to model the laws, a scientist could predict the future behaviour of the world, and retrodict how it behaved in the remote past.

Of course, there is more to the world than falling bodies, and the full impact of these revolutionary new ideas had to await Newton's monumental work in the late seventeenth century. Newton went beyond Galileo and developed in detail a comprehensive system of mechanics, capable of dealing in principle with motion of all types – and it worked. The new scope in physics demanded new advances in mathematics to describe the laws that Newton discovered. The so-called differential and integral calculus was invented. Once again, time played a central role in stimulating these developments. How rapidly would a body change its speed under the action of a certain force? How fast would the force vary as its source moved about? These were the kind of questions which the new mathematics had to answer. Newton's mechanics is a description of *change*, the reorganization of the world according to the passage of time.

As a result of this reorientation in thinking, new types of questions were asked about the universe in which time and change formed a prominent part. Whereas in the older cultures, balance and equilibrium – features so relevant to the well-being of biological organisms – were the important issues, Newtonian mechanics emphasized the dynamic aspects of nature. It is perhaps no coincidence that in spite of the explosive growth of civilization in classical times, pre-Renaissance cultures were largely static, seeking to maintain the status quo. By contrast, Galileo and Newton, and later Darwin, introduced the crucial concept of evolution to mankind's view of nature. As so often in the development of human thought, it is a change in the perspective, rather than a new piece of information, which leads to intellectual revolutions. Older cultures had been concerned with such issues as how to avoid the displeasure of the storm god and ensure a good harvest but Newton and his mathematics addressed a totally new type of problem: given the present state of a physical system, how will it develop in the future? What final state will result from a given set of initial conditions?

These intellectual developments were accompanied by social changes: the industrial revolution, the systematic search for new knowledge and technology, and above all the concept – so much taken for granted today – of a community *progressing* towards a better standard of living and control over its environment. The transition from a static society influenced by temperamental nature, to a dynamic one seeking to control nature, owes much to the new mechanics with its crucial concept of temporal evolution.

Another important idea that became properly clarified by Newtonian mechanics is that of alternative futures, a concept central to the theme of this book. To understand the implications requires a careful consideration of just what is meant by a mathematical law of nature. As we know, Galileo and Newton discovered that the motion of material bodies is not a random and haphazard affair, but one restricted by simple mathematics. Thus, given information about the state of a body and its surroundings at one instant, it is possible (in principle at least) to compute the behaviour of the body in the future (and past). Careful experiment confirms that this is correct. The whole spirit of this idea is that the world cannot change in any way; the available paths of development are constrained to those that conform to the laws. But just how restrictive is this constraint? A straightjacketed world does not interweave easily with our experience of nature, full of a rich and seemingly limitless variety of interesting and complex activity.

The reconciliation of complexity and compliance is found in the form of the mathematics needed, and its relation to the requirement for 'information' about the system at some initial moment. To make this precise we could consider the very simple practical matter of throwing a ball. Newton taught us that the path of a missile is not arbitrary, but must be a well-defined curve shaped in accordance with his mathematical laws. Nevertheless it would be a boring world for sportsmen if all projected balls followed exactly the same trajectory, and of course we know this does not happen. In fact, the laws do not define a unique trajectory at all, but only a class of trajectories. In the case under consideration, every ball will follow a parabolic path, but there are an infinite variety of parabolas. (A parobola is the shape you get by slicing through a cone parallel to the opposite face. The parabola itself is the shape of the curved edge of the truncated cone.) There are high, thin parabolas corresponding to balls thrown almost vertically, long low parabolas like the path of a baseball, and so on. In fact, experience shows that we have control over the shape of the path in two ways. We can decide the size of the parabola by varying how fast we throw the ball, and we can vary the shape of the parabola by adjusting the angle of projection. Thus, it is a law of physics that all thrown balls follow parabolic paths, but which parabola is determined by two independent initial conditions: speed and angle.

The purpose of this digression into elementary ballistics is to point out that there is more to nature than laws alone. There are also initial conditions. We can now clarify the question of what information is required to determine the particular behaviour of a body according to Newtonian mechanics. First, one needs to know the magnitude and direction of all the forces which act on the body, and how they vary with time, and secondly the position and velocity of the body at some moment must also be specified. Given all this data, it is then just a matter of mathematics to calculate where the body will be, and how it will be moving, at some later moment.

One of the early successes of his mechanics was Newton's explanation of the sizes, shapes and periods of the planetary orbits in the solar system. The planets, Earth included, are trapped in orbit about the sun by the gravity of this body. To compute the motion of the solar system, Newton had to know both the strength and direction of the sun's gravitational force everywhere throughout space, and also the initial conditions, i.e. the positions and velocities of the planets at a particular moment. The latter information could be supplied by the astronomers, who routinely monitor such matters, but the strength of gravity was quite another issue. Generalizing Galileo's results concerning the Earth's gravity, Newton correctly guessed that the sun, and indeed all bodies in the universe, exert a gravitational force which diminishes with distance according to another precise and simple mathematical law; the so-called inverse square law. Having mathematized motion, Newton also mathematized gravity. Putting the two together and using calculation led to a great triumph when he correctly predicted the behaviour of the planets.

Since Newton's time, his mechanics have been applied to the solar system in very great detail. It is possible to improve on the original calculation by taking into account minute forces of gravity produced between the planets themselves, by the effects of their rotation, distortions in their shapes and so forth. A standard calculation is to compute the orbit of the moon and thereby predict the dates of forthcoming eclipses. Similarly, the calculation can be applied backwards to determine the dates of past eclipses and check them against historical records.

The application of Newtonian mechanics to the solar system was more than just an exercise. It exploded centuries of belief that the heavens were ruled by purely celestial forces. But even the great refuge of the gods succumbed to Newton's mathematics. Never was there a more dramatic demonstration of the power of science based on mathematical law. It implied that laws of nature not only control minor processes on Earth, such as the shape of projectile trajectories, but also rule the very structure of the cosmos, a widening in perspective to the cosmic domain which profoundly altered mankind's conception of the nature of the universe and his place within it.

The deep philosophical implications of the Newtonian revolution are most sharply focused by the subject of cosmology: the study of the totality of things. According to Newton, the motion of every speck of matter, every atom, is in principle completely and absolutely determined for all of past and future time by a knowledge of the impressed forces and the initial conditions. But the forces themselves are in turn determined by the location and condition of matter. For example, the sun's gravitational force is fixed once we know its position. It follows that once we know the positions and motions of all the bits of matter, and assuming we know all the laws governing the forces between the bits, then we can compute the entire history of the universe, a possibility pointed out by Pierre Laplace.

Now it must be said right away that this knowledge is not available, and even if it were, no computer would be large enough to perform the calculation. In practice, of course, only very simple, relatively isolated subsystems (e.g. the solar system) can be handled computationally. Nevertheless as an issue of principle it is still arresting in its implications. The ancient idea of the cosmos as a society of temperaments, coexisting in equilibrium, has given way to an inanimate, even sterile, picture of a *clockwork universe*. Newton's discoveries seem inevitably to relegate the entire world to the status of a mechanism, inexorably and systematically ticking its way forward to a prearranged destiny, with every atom careering along some convoluted, but legislated, pathway to an unalterable destination.

The shift in perspective eventually made its impact on religion. The early Christian idea of an active God closely involved in the affairs of the world, supervising events from the conception of a child to the phases of the moon, gave way to the more remote idea of God as initiator of the cosmic action, passively watching His creation unfold according to His mathematical laws. God had been transformed from architect to mathematician. The spirit of this transformation to divine passivity and unsupervised lawfulness is captured in Robert Browning's poem 'Pippa Passes': 'God's in His heaven, All's right with the world'. The clockwork universe, smoothly developing according to plan, had arrived: such was the impact of Newton's genius small wonder that Pope should write 'God said "Let Newton be!" and all was light'.

In spite of the startling intellectual achievement in bringing discipline to an unruly cosmos, the establishment by Newton of a law-abiding universe has a profoundly depressing aspect to it. When the last atom has been brought into line, as it were, some spark of life goes out of the world. A clockwork mechanism can be very beautiful and efficient, but the image of a universe brainlessly cavorting its way to eternity like some grotesquely complicated musical box is not especially reassuring, particularly as we ourselves are part of it. One obvious casualty is free will. If the entire past and future condition of all matter is uniquely determined by its condition at any one instant, then our future must obviously be predetermined in every last detail. Every decision we make, every random whim, must in reality have been arranged billions of years in advance to be the inevitable outcome of a staggeringly intricate but fully determined network of forces and influences.

Today, scientists recognize several flaws in the argument leading to a predetermined, clockwork universe, but even granted the essential idea it must not be supposed that Newtonian laws are so restrictive that they permit only one possible universe. There are still the initial conditions. Just as a ball may follow any one of an infinite variety of trajectories, so may the entire universe follow an infinite variety of pathways into the future. Precisely which pathway it has chosen is determined by the initial conditions. This raises the central question of what is meant by 'initial'. Later we shall see that modern cosmologists believe the universe did not always exist, so that there must have been some sort of creation, thought to have occurred about fifteen billion years ago. Thus, we may meaningfully ponder the following fascinating problems. What initial conditions at the creation led to the universe we now see? Were these conditions very special and contrived, or were they typical of a wide class of conditions? What sort of universe would we have ended up with had the conditions been different?

The underlying philosophy here is that our universe is just one of an infinite set of possible universes – simply one particular pathway to the future. We can study the other pathways using mathematics. We can probe the nature of these myriad alternative worlds that might have been and ask: why this one? In the coming chapters we will see how closely our own existence is involved in these issues, and how these other ghost worlds are not merely academic curiosities but can actually make their presence felt in the concrete world of our experience.

One of the oddities about a Newtonian mechanistic universe is its apparent contradiction with experience. So much of the world about us seems to happen by chance rather than design. Contrast, for example, the behaviour of a thrown ball with that of a tossed coin.

Both move according to the principles of Newtonian mechanics. The ball, if thrown repeatedly at the same speed in the same direction will always follow the same path, but the tossed coin comes down sometimes heads, sometimes tails. How can these differences be reconciled with the world as a fully determined sequence of events?

First consider what is meant by a law of nature. As conceived by classical scholars, and later incorporated into Newton's conception of mechanics, a law was supposed to describe how a particular physical system would behave under a particular set of circumstances. As natural laws are, by definition, not meant to change from time to time or place to place, it is clear that they are closely connected with repeatability, a concept central to the philosophy of verifying theories by repeated experiment. Consequently, if a thrown ball moves according to Newton's laws, and if the ball is thrown and re-thrown under duplicate conditions, then its path must each time be the same.

A good way of analyzing this problem is using the notion, introduced above, of a whole collection of worlds, often called an ensemble. Imagine an ensemble (infinite if you like) of worlds, identical except for the career of the ball. In each world the ball is projected at a slightly different angle and/or speed. There are a whole series of paths – one for each world; all are parabolic, but no two paths are identical. It is helpful to have some way of labelling the worlds to distinguish between them. One useful method is to draw a diagram in which the two initial conditions – speed and angle – are plotted together (see Figure 1). Each pair of numbers (speed, angle) defines a point in the diagram, and corresponds uniquely to one particular world and one particular trajectory. Thus, the worlds are labelled by the number pairs.

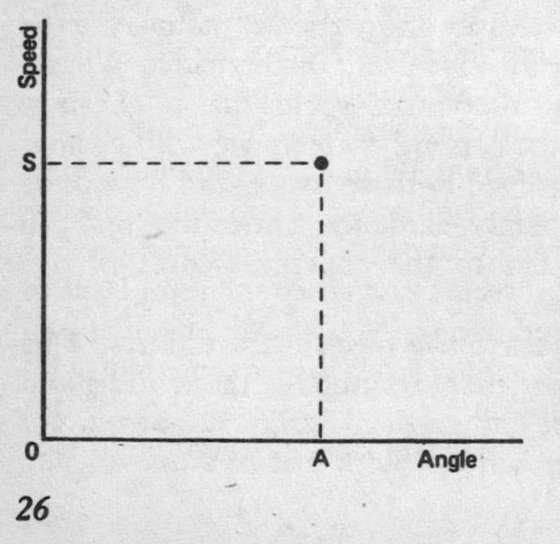

Figure 1
The point on the diagram fixes a particular speed S and angle A. With these two initial conditions the shape of the ball's path is uniquely defined by Newton's laws of motion.

Now consider a whole family of other points surrounding the one of interest (Figure 2). These points represent other worlds which are, in some sense, close neighbours of the original. They represent worlds in which the initial conditions have been very slightly disturbed. If we ask about the behaviour of the balls in these neighbouring worlds, we find that their paths are all very close to that of the original. In short, a small change in the initial conditions causes a small change in the subsequent motion.

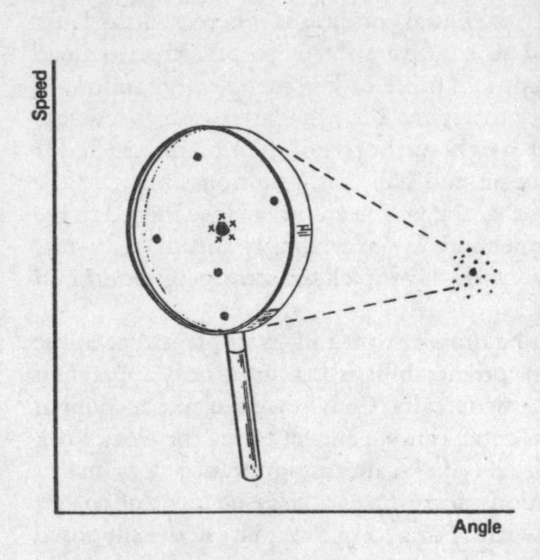

Figure 2
The cluster of twenty nearby points could represent twenty worlds which differ only in the slight variations of a ball's parabolic path. Alternatively, they could represent worlds in which snooker balls end up in widely different arrangements. Only in the magnifier do we notice that the latter process is not inherently random, but merely very sensitive to small changes in the initial motion of the cue ball. So there do actually exist points (marked by crosses) very close indeed to the original point, which would lead to almost identical snooker ball configurations.

In contrast to this, consider another familiar situation involving balls, this time several of them. In the game of snooker, play begins when one of the players projects the cue ball at a group of ten others closely packed in the shape of an inverted triangle. After impact, the balls scatter about the table, colliding and bouncing off the sides, until finally they all come to rest (due to friction) in some configuration. No matter how many times we care to repeat the procedure, and however carefully we line up the cue ball, it seems that we could never hope to produce exactly the same configuration twice. The outcome is apparently neither predictable nor repeatable. How is this consistent with deterministic Newtonian mechanics?

Returning to Figures 1 and 2, we may still label the members of our ensemble of worlds by points in the diagram, because given a single point, i.e. a particular angle and speed of the cue ball, the end configuration of balls is completely determined by the laws. The difference here from the case of the single thrown ball lies in the properties of the ensemble, not in any single world, for even initial conditions which are very close indeed to the original will produce drastically different end configurations of balls. Any minute change in speed or angle will scatter the balls in a completely different way.

These two cases can best be contrasted by saying that in the former we have good control over the initial conditions whereas in the latter we do not. The snooker ball configuration is so sensitive to small perturbations that the outcome is more or less completely random. If we could apply a magnifier to Figure 2 for the latter case, we would notice that in fact there are neighbourhoods of each point which, for that world, would produce an end ball configuration similar to the first break. The problem is that the points are very close indeed to the first point, i.e. the neighbourhoods are exceedingly shrunken, so that for all practical purposes we could never pick the same neighbourhood twice.

The conclusion that can be drawn from this example is that in the real world, the deterministic predictability of nature is only apparent if we can view the world microscopically. Only if we can take account of the career of every atom in detail can we expect to see the clockwork mechanism at work. On an everyday scale, our ignorance of, or lack of control over, initial conditions, introduces a large element of chance into the behaviour of the world. For a long time, physicists supposed that this purely practical limitation was the only source of uncertainty and chance. The atoms themselves were supposed to move according to the deterministic laws of Newtonian mechanics, that is, atoms were thought to differ from macroscopic objects such as billiard balls only in the quality of scale. Indeed on this assumption physicists were able to account satisfactorily for many of the properties of gases and solids, by viewing them as vast accumulations of atoms, each moving according to Newton's laws. Of course, some averaging procedure had to be adopted as the actual motions of the individual atoms could not be calculated in practice. However, the gross behaviour of all the atoms could be predicted.

About the turn of the century it was discovered that atoms are not solid, indestructible bodies after all, but possess an internal structure, rather like the solar system, with a heavy nucleus at the centre surrounded by a cloud of light, mobile electrons. The whole system is bound together by electric forces which attract the negative electrons to the positive nucleus. It was natural that physicists should look to Newtonian mechanics as a mathematical model of the atom, in an attempt to repeat the earlier success in explaining the motions of the solar system. Unfortunately, the model seemed to contain a fundamental flaw. It was discovered in the nineteenth century that when an electric charge accelerates it emits electromagnetic radiation, such as light, heat or radio waves. A radio transmitter uses this principle by forcing electrons up and down an aerial. In an atom also, electrons are forced into curved orbits by the electric field of the nucleus, and this acceleration should cause them to radiate. If this happens, then the system loses energy to the radiation, and to pay for it the atom must shrink. Thus, the electron is drawn closer to the nucleus, and has to orbit still faster to overcome the stronger electric field there. The result is an even more energetic emission of radiation and a faster shrinkage. In fact, the system is unstable, and the atom should collapse completely after a very short time. What is wrong?

The answer to this enigma was not fully discovered until the 1920s, although some faltering steps were taken as early as 1913. In later chapters the resolution will be considered in more detail, but for the moment suffice it to say that not only did Newton's laws fail to apply for atoms, but so did any other laws of the type known hitherto. The replacement theory not only demolished two centuries of science, but challenged some fundamental assumptions about the whole meaning of matter and our observation of it. This quantum theory, as it is now called, was developed in stages from 1900 until about 1930. It has the most profound consequences for the nature of the universe and our place in it.

The experiments conducted by Davisson, mentioned at the beginning of this chapter, were the first direct observation of the astonishing new principles at work. As an introduction to the new theory, let us re-examine the idea of a law of motion. Suppose a ball is projected from a place A and moves along a path to some other place B. If the procedure is repeated, we expect the ball to follow exactly the same path (so long as the initial conditions are identical). This property was also expected of atoms, and their constituent particles, electrons and nuclei. The shattering discovery of the quantum theory was that this is not so. A thousand different electrons will travel from A to B along a thousand different paths. The rule of mathematical law over the behaviour of matter appears at first to be finished, presenting the spectre of subatomic anarchy. It is hard to overemphasize the immense implications of this discovery, for ever since Newton had found that matter behaves according to definite rules, it was supposed that some sort of rules would apply on all scales, from the atom to the cosmos. But now it seems that the orderly discipline in the macroscopic world of our experience collapses into chaos within the atom.

Although, as we will discover, subatomic chaos of a sort is inescapable, this chaos can, by its very nature, produce some form of order. To clarify this rather enigmatic statement, consider a park with a fence around it and two gates on opposite sides, labelled A and B. Suppose the park is situated on some frequently used thoroughfare, so that people tend to enter by gate A, stroll across to B, and exit. If we were to plot the tracks of the visitors to the park in, say, one hour, we should end up with a diagram like Figure 3. The characteristic feature is that

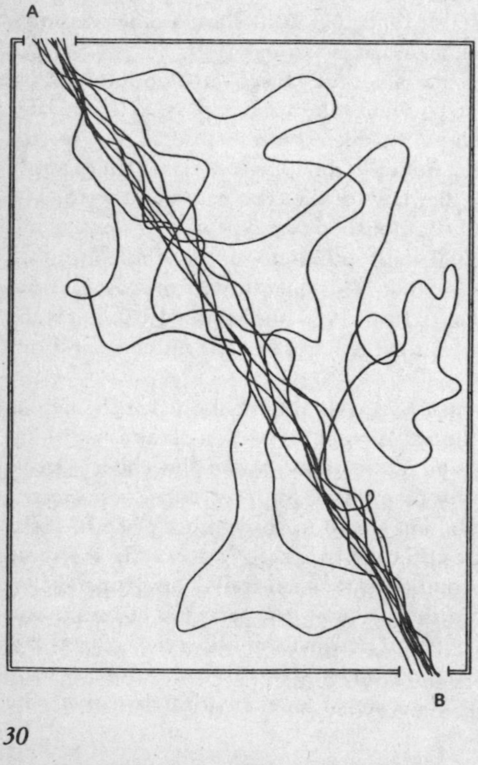

A

B

Figure 3
Paths through a park. Most people try to minimize their activity and walk the shortest route, so the ensemble of paths is concentrated along a straight line between entrance and exit. Some energetic folk, however, execute quite complicated perambulations. Subatomic particles also follow a multiplicity of routes, but prefer the short ones.

most of the visitors walk fairly close to a straight line between A and B. Some, with more time or energy, wander a little way to one side, while a few (perhaps those with dogs or still more energy) stroll around near the edges of the park. Occasionally there will be a very haphazard path (a child?). The point is that people are not apparently subject to any rigid law of motion; they regard themselves as free to choose any path through the park. Indeed, any individual may decide to stray quite a way from the shortest route. In spite of this, when a sufficiently large group is studied, it is highly probable that there will be a concentration of paths near the straight one. Given enough subjects, some sort of order emerges, even though the law 'walk straight' is generally violated. The reason is that when a large collection of people are considered, various individual whims and fancies get averaged out, and the collective behaviour displays unwitting conformity. The reason behind the particular conformity discussed here is that people, on average, are more likely to choose low action behaviour than to involve themselves in a high level of activity. The straight path from A to B is the path of least action, and hence the one most likely to be followed by any given pedestrian. But it does not *have* to be so; it is all a matter of probabilities.

The example of the strollers in the park is very similar to that of the subatomic particles, which also choose a whole variety of paths from A to B, though the ones they prefer are the low action ones. So once again the paths tend to cluster round the path of least action. Electrons it seems, like humans, do not want to exert themselves too much. Now the significant fact about the path of least action is that it coincides with the Newtonian path – the trajectory one would calculate from Newton's laws.

Returning to the example of the strollers in the park, we could also observe another interesting feature. Fat, heavy people are more likely to follow the straight route than small, light people (e.g. children). This is because the extra action required to move a heavy body along a winding path is greater than for a light one. Similarly for particles of inanimate matter: the heavy ones, such as atoms, or whole groups of atoms, are more likely to follow close to the least action path than electrons. When the particles are so heavy that they become macroscopic (e.g. billiard balls) then they are exceedingly unlikely to roam more than an infinitesimal distance from the Newtonian path of least action. We can now understand how atomic anarchy is consistent with Newtonian discipline where everyday bodies are concerned. Deviations from the law are allowed, but are utterly minute except on subatomic scales, so we do not normally notice them.

Using a mathematical principle comparable to the human quality of reluctance to engage in unnecessary activity, the quantum theory allows one to compute the relative probabilities of all the different paths that an electron, or an atom, may follow. Basically the action required for a particle to move along a given path is calculated (this requires a precise definition of 'action') and inserted into a mathematical formula that supplies the probability for that path. All paths are generally possible, but not all are equally probable.

We still need to know how all this prevents atoms from collapsing. A further startling revelation about the nature of subatomic matter, which will be deferred till chapter 3, is also necessary, but for now a rough idea can be given. According to the old theory a particle orbiting a nucleus should steadily spiral inwards as it divests its energy in the form of electromagnetic radiation. This is the classical path. However, quantum theory allows many other paths to be followed. If the atom has a lot of internal energy, then the electron will reside far from the nucleus and its behaviour will not depart too much from the classical picture. However, when a certain amount of energy has been lost by radiation, and the electron drops closer to the nucleus, a new phenomenon occurs. It is important to remember that the electron is not just travelling along a simple path from A to B, but is orbiting round and round. Thus, the possible paths cross and recross in a complicated way, a feature that must be allowed for when computing the most probable behaviour of the electron. It turns out to have crucial significance: there is a state of minimum energy below which the probability of finding an electron is strictly zero. In its motion the electron can make temporary excursions towards the nucleus, but it is forbidden to stay there. The average location of the electron works out at about ten billionths of a centimetre from the nucleus, which is the radius of an atom in its lowest energy state.

There are in fact a whole sequence of these energy levels for the atom, and light is emitted when the electron makes a downward transition from one energy level to another. Because the levels represent a fixed energy, the atom will not just emit any amount of light, but pulses or packets containing a given quantity of energy, characteristic for each type of atom. These packets of energy are called quanta, and quanta of light are known as photons. The existence of photons was known long before the atomic theory was finally worked out as described here: the work of Planck, together with Einstein's explanation of the photoelectric effect, showed that light only comes in discrete units of energy. The energy of each of these photons is proportional to its frequency, so that the colour quality of light is a measure of its energy. Thus blue light, which is high frequency, has rather more energetic photons than a low-frequency colour such as red. Furthermore because a certain type of atom (e.g. hydrogen) will only emit certain quanta, the quality of light from each species of atom will be a distinctive label. For the colours of light from hydrogen differ completely from the colours from, say, carbon. Of course, each atom can emit a whole range, or spectrum, of colours, corresponding to the entire sequence of energy levels (they are not equally spaced in energy), and in this way quantum theory can be used to explain the characteristic light spectra of different chemicals. Indeed computations can be performed that give not only the exact colours, but their relative strengths, by computing the relative probabilities for electrons to follow the various jumpy paths between the different levels.

These sweeping successes of the quantum theory are impressive enough, but they are only the beginning. In later chapters we shall see how the applications range far wider than atomic structure and spectra. One thing, though, has not yet been properly explained: exactly how the crossing and recrossing of the paths leads to such drastic changes in the behaviour of the electrons. There is a deep mystery here. How does an electron *know* it has crossed its own path? There is a still more extraordinary phenomenon which will be discussed in chapter 3: the electron not only has to know about its own path, it must also know about all the other paths it never actually follows!

To summarize the significant features of the quantum revolution, we find that rigid laws of motion are really a myth. Matter is allowed to roam more or less at random, subject to certain pressures, such as the reluctance to engage in too much activity. Complete chaos is thus averted because matter is lazy as well as undisciplined, so that in a sense the universe avoids total disintegration because of the inherent indolence of nature. While no definite statement can be made about any particular motion, certain paths of behaviour are more probable than others, so that statistically we can predict accurately how a large collection of similar systems will behave. Although these strange features are only pronounced on the atomic scale, it is clear that the universe is not, after all, a clockwork machine whose future is completely determined. The world is ruled less by rigid laws than by chance. The uncertainties, moreover, are not merely a result of our ignorance of initial conditions, as was once thought, but an inherent property of matter. So unpalatable did this inherent chanciness of nature seem to Albert Einstein that he refused to believe it throughout his life, dismissing the idea with the famous retort 'God does not play dice'. This notwithstanding, the vast majority of physicists have come to accept it. Following chapters will reveal the amazing consequences of a fundamentally uncertain cosmos.

2. *Things are not always what they seem*

In the last chapter we learnt just how central to our perspective of the world is Newton's idea of a mathematically precise time, flowing uniformly and universally from past to future. We do not see the world as static, but as evolving, developing, changing from one moment to the next. It was once thought that the future states of the world unfolded in a way which was predetermined by its present state, but the quantum revolution demolished that. Instead the future is inherently uncertain. The quantum theory tumbled the edifice of Newton's mechanics, but what of his model of space and time? This too has collapsed, in a revolution equally as profound as the quantum theory, but predating it by several years.

In 1905 Albert Einstein published a new theory of space, time and motion called special relativity. It challenges some of the most cherished assumptions normally made about the nature of space and time. Since the original publication, the theory has been verified repeatedly in laboratory experiments and is accepted almost unanimously by today's physicists. Among the more spectacular predictions of the theory are the existence of antimatter and time travel, the elasticity of space and time, the equivalence of mass and energy and the creation and annihilation of matter. In an extension to his 1905 work, Einstein published the so-called general theory of relativity in 1915. Although not so well founded experimentally, its predictions are just as bizarre: curved space and time, black holes, the possibility of a finite but unbounded universe, and even the possibility of space and time smashing themselves out of existence.

The theory of relativity embarks upon these extraordinary possibilities by adopting a radically new perspective of exactly what the world *is*. According to Newtonian ideas – being the common-sense perspective adopted by ordinary people in everyday life – the world changes from moment to moment. At any given moment, *the* world is some well-defined (though not completely known) state of the whole universe. We inevitably think of all the other people, the other planets and stars, the other galaxies – whatever interests us – and envisage them in some particular condition at this moment, i.e. now. The world is therefore seen as the totality of all these objects at one particular time. Most people do not doubt the existence of a universal 'same moment' (Newton didn't).

The demise of this familiar way of thinking about time is revealed by a curious phenomenon. Between the constellations of Aquila and Sagitta is a weird astronomical object known as a binary pulsar. It apparently consists of two imploded, or collapsed, stars in close orbit about one another. The stars are believed to be so compact that even their atoms have collapsed into neutrons under their own weight in the intense gravity. Because of the enormous compaction – the stars are barely a few miles across – they can rotate tremendously fast, several times a second. One of the stars is evidently surrounded by a magnetic field, for every time it rotates it emits a pulse of radio waves (hence the name pulsar), and for the last few years astronomers have been monitoring these blips from the giant radio telescope at Arecibo in Puerto Rico. The regularity of the neutron star's rotation is reflected in the precise regularity of the pulses, which can therefore be used as an accurate star clock, as well as allowing the motion of the star to be followed.

The regularity of the pulses provides a graphic illustration of the inadequacy of common-sense time. Being so massive and so close together, the two neutron stars dance around each other at phenomenal speed, taking only eight hours to complete one orbital revolution – an eight hour 'year'. The pulsar therefore moves at a good fraction of the speed of light, which is the same as the speed of the radio pulses. (Light, radio waves and other radiations such as infra-red heat, ultra-violet, x and gamma rays are all examples of the same basic phenomena – electromagnetic waves). As the pulsar orbits round its companion it sometimes approaches the Earth and sometimes recedes, depending on its momentary direction of motion. Commonsense would suggest that when the pulsar approaches its radio pulses are speeded up, because they receive an extra push in our direction from the motion of the star itself, rather like a sling shot. Similarly, the pulses from the pulsar when it is receding should be slowed down. This being the case, the former sequence of pulses should arrive long before the latter because they will cover the enormous intervening distance to Earth at a higher speed. Indeed, the pulse arrivals for the whole orbit should be scrambled up over an interval of many years, thus mixing together pulses from thousands of orbits in a complicated muddle. Observations, however, show something quite different: a regular pattern in which the pulses from all the orbital positions arrive arranged neatly in the correct sequence.

The conclusion seems enigmatic: there are no fast-moving pulses overtaking slow-moving pulses. They all arrive at the same speed, spaced out between each other in a regular way. This seems in direct contradiction to the fact that the pulsar is moving, and a vivid demonstration of the contradiction is provided by the fact that the same pulses that arrive with unaltered speed also carry direct information that the pulsar really is rapidly moving. The information of interest is encoded in the quality of the radio pulses themselves, which are of a longer frequency when the pulsar is receding than when it is approaching. This frequency shift, similar to the change in engine pitch when a fast car speeds by, is used by police on radar traps to measure the motions of cars. The same technique shows the pulsar hurtling through space, and yet its pulses reach Earth with unchanged speed.

A century ago observations such as these would have caused consternation, but today they are expected. As long ago as 1905, Einstein predicted such effects on the basis of his theory of relativity. A combination of mathematical theory and experiment led Einstein to a remarkable – indeed, scarcely believable – conclusion: the speed of light is the same everywhere for everybody, and this is true no matter how they are moving. In those days the motivation behind this cryptic pronouncement had to do with the properties of moving electric charges and the inability of physicists to measure the speed of the Earth by using light signals. We shall not dwell upon the technical details here, except to say that *the* speed of the Earth turns out to be completely meaningless, as only relative speeds (hence the appellation 'relativity') can ever be measured. Instead, let us concentrate on the meaning and implications of Einstein's pregnant suggestion.

If an object is receding from you, and you start to chase after it, you expect this manoeuvre to result in a diminished rate of recession. Indeed if enough effort is put into the chase, you may even succeed in overtaking the object. Thus, the relative speed between you and it clearly depends on your state of motion. If, however, the object is a pulse of light, this is not so. Incredible though it may seem, no matter how much effort you put into the chase you never gain one single mile per hour on the light pulse. True, light moves very fast (186,000 miles every second) but even if you could travel in a rocket at 99.9 per cent of the speed of light, you would never succeed in reducing its rate of recession, however powerful the rocket motors.

Now these statements probably seem like sheer nonsense. If someone remaining on Earth watches the chase, and they too see the light pulse receding at 186,000 miles per second, with the rocket plunging after it almost as fast, they *must* see the gap widening at only a fraction of the speed of light. Yet, if Einstein's proposal is accepted (and experiment confirms it is correct) then the rocket man sees the same light pulse opening up a lead at the full 186,000 miles per second. The only way to reconcile these apparently conflicting observations is to suppose that the world from the rocket looks and behaves very differently from the world as viewed from Earth.

One striking demonstration of this difference appears if the astronaut conducts an experiment with light pulses inside the spacecraft at the moment he passes by his Earthbound colleague (Figure 4). At this instant he arranges for two pulses of light to be sent in opposite directions from the exact centre of the rocket, one towards the front, the other towards the rear. Naturally he sees both pulses strike the opposite ends of the rocket simultaneously. Remember that the immense forward speed of the rocket relative to Earth has no effect whatever on the speed of the light pulses as observed from the rocket. However, these events as witnessed from the Earth cannot be the same. During the brief interval of time taken for the pulses to travel the length of the rocket, the rocket itself moves forward appreciably. The Earthbound observer also sees the two pulses travel at equal speeds relative to *him,* but from his frame of reference the rocket is in motion – the front end of the rocket appears to be retreating from its pulse and the rear end advancing to meet its pulse. The inevitable result is that the rear pulse arrives first. The two events are not simultaneous as observed from Earth, but are simultaneous when viewed from the rocket. Who is correct?

The answer is that both are correct. The concept of simultaneity –

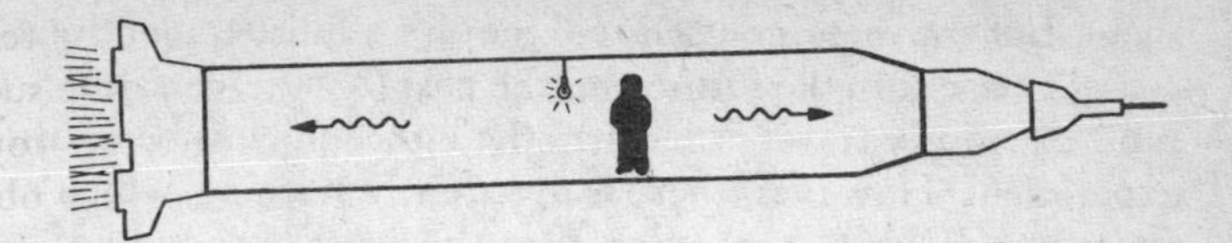

Figure 4 There is no universal present
This startling conclusion follows from the peculiar behaviour of light pulses. In the spacecraft the pulses strike the end bulkheads simultaneously because they travel at equal speed (as viewed from the rocket) from the centre of the craft. Observed from Earth, however, the pulses also appear to travel at equal speed, so the left-hand pulse arrives first because the rear bulkhead is plunging forward to meet it while the front bulkhead is retreating.

the same moment in two different places – has no universal meaning. What is judged to be 'now' by one observer can be in the past or future as determined by another. At first sight, such a conclusion seems alarming. If one person's present is another's past and yet another's future, couldn't they signal each other and enable the future to be foretold? What would then happen if the informed observer acted to change this already-observed future? Fortunately for the consistency of physics, it does not seem that this situation can occur. For example, in the case of the rocket experiment, the observers can only know that the light pulses have arrived when some sort of message gets back to them. But the message itself will require a finite travel time. To beat causality and turn the future into the past (or vice versa), this message must obviously travel faster than the light which is being used in the experiment. However, there is apparently nothing which can travel faster than light. If there were, then the causal structure of the world would be threatened. So we see that 'pastness' and 'futurity' are not really universal concepts, but only apply to events which can be connected by light signals.

We might well wonder why a rocket cannot simply go on accelerating, and thereby be observed, from Earth, to overtake light. Einstein showed that this is impossible. As the light barrier is approached the rocket and occupants begin to grow heavier and heavier. More and more energy must be expended in overcoming the extra inertia to shift them faster. The gain in speed steadily diminishes, and the speed of light itself is never reached, however long one waits. Naturally the astronaut does not regard himself as putting on weight; instead the world around him appears strangely distorted. Crudely speaking, distances in the forward direction seem shrunken. Consequently, viewed from the rocket, the astronaut *does* appear to be going faster and faster, because he seems to have less and less distance to travel in a given time.

An astronaut in a rocket travelling at 99.9 per cent of light speed would regard the sun as a mere four million miles from Earth, and he would reach it in only 22 seconds. Incredibly though, observers on Earth, who do not perceive this shrunken perspective, measure the distance to the sun to be 93 million miles and the duration of this much longer journey to take over eight minutes. The conclusion seems to be that time as observed from the rocket runs twenty-two times slower than it does on Earth. The real surprise, though, comes when the astronaut looks back at the Earth. If events in the rocket really do occur twenty-two times slower than events on Earth, then it might appear that if the astronaut could look back at Earth through a telescope, he would see things running twenty-two times as fast as normal. In fact, instead of seeing events speeded up twenty-two times, he sees just the reverse – a slow-motion Earth. Evidently *both* observers regard the other's time as running slow. This symmetric relationship between moving observers lies at the heart of the theory of relativity, which assigns meaning to motion only relative to other observers. It is, therefore, impossible to say the rocket is moving and the Earth still, or vice versa, so any effect witnessed by one must be witnessed by the other also. There is no real inconsistency in the fact that each observer sees the other's time slowed up when we recall that they disagree greatly on which moment in the other's frame of reference is to be regarded as the 'present'. They can only compare times by the lengthy process of sending signals backwards and forwards, which takes as least as long as the light travel time between them.

The reality of the time dilation effect becomes manifest if the rocket returns to Earth, and a comparison is performed directly between the clocks on Earth and that in the rocket. The astonishing discovery is that the two observers' times have been wrenched permanently out of step. What may have been a few hours' journey to the astronaut will have occupied days of Earth time. Nor is this just a strange physiological effect: the rocket *will* have experienced only a few hours duration in several days duration as experienced on Earth.

The concept of elastic time was quite a shock when Einstein introduced it in 1905, but since then many experiments have confirmed its reality. The most accurate of these uses subatomic particles because they are easy to accelerate to near the speed of light, and they often contain an inbuilt clock. Mu-mesons, or muons for short, can be created in controlled subatomic collisions, and have a lifetime of around two microseconds, before disintegrating into more familiar particles of matter, such as electrons. When moving at close to the speed of light, the dilation of time increases their lifetime as measured by us, by several times. Of course, in their own frame of reference they still live for two microseconds. A good check on the effect was made at the particle accelerator laboratory at CERN in Geneva in early 1977, where a beam of high speed muons was created and stored in a magnetic ring so that their lifetime could be measured. It confirmed the amount of time dilation predicted by the theory of relativity, to an accuracy of 0.2 per cent.

One intriguing possibility opened up by the time dilation effect is time travel. By approaching closer and closer to the speed of light, an astronaut can dislocate his time scale more and more violently relative to the rest of the universe. For example, rocketing to within one hundred mph of the speed of light, he could accomplish a journey to the nearest star (more than four light years away) in less than one day though the same journey measured from Earth takes over four years. His clock rate is thus about 1800 times slower when observed from Earth than when observed from the rocket. At only one mph less than light speed, the dilation is increased to 18,000 times, and the journey from the rocket seems like a bus ride, although still several years whilst being watched from Earth. At this colossal speed, the astronaut could circumnavigate the entire galaxy in a few years (rocket-time) and return to Earth to find himself in the four thousandth century! Although such feats of travel may remain forever in the realm of science fiction (it would consume enough energy to power all our present technology for millions of years) time dilation is definitely science fact.

The purpose of mentioning these bizarre effects is to emphasize that notions like space and time are not as concrete or universal as most people think. The essential element injected into physics by the theory of relativity is subjectivity. Fundamental things like duration, length, past, present and future can no longer be regarded as a dependable framework within which to live our lives. Instead they are flexible, elastic qualities, and their values depend on precisely who is measuring them. In this sense the observer is beginning to play a rather central role in the nature of the world. It has become meaningless to ask whose clock is 'really' right, or what is the 'real' distance between two places, or what is happening on Mars 'now'. There is no 'real' duration, extension or common present.

At the beginning of this chapter we found that relativity adopts a totally new perspective on what the world 'really' is. In the old Newtonian picture, the universe consists of a collection of *things*, located here and in other places at this moment. Relativity, on the other hand, reveals that 'things' are not always what they seem, while places and moments are subject to re-interpretation. The relativist's picture of reality is a world consisting of *events* rather than things. Events are points of space and time without extension or duration: five o'clock at the exact centre of Piccadilly Circus is an event (though possibly a rather uninteresting one). Events are universally agreed by all observers, though there will generally be disagreement about where or when the events occurred.

In spite of the relativity of what were formally regarded as concrete, absolute qualities, some common sense organization of space and time remains. For example, the discrepancies between the 'present moment' as interpreted by different observers, and the elastic stretching of time, cannot become so violent that it actually flips past into future in a way that a single observer can see. That is, although some events may be regarded as past to one observer, future to another and present to a third, a sequence of two causally connected events will always be witnessed in the same order. If the firing gun causes the target to smash, then no observer, irrespective of his state of motion, will see the target smash before the gun is fired. The correct causal relationship is only maintained, though, because of the rule that observers cannot break the light barrier and travel at superluminal speeds. If this were possible, cause and effect could be interchanged, and an astronaut could travel backwards in time as well as into the future. We should then be presented with a fate similar to that of Miss Bright, who

> could travel much faster than light.
> She went off one day, in a relative way,
> and came back the previous night.

The causal chaos threatened by visiting one's own past seems to be a fictional possibility only.

In a world of shifting space-time perspectives, a new language and geometry is required which takes into account the observer in a fundamental way. Newton's concepts of space and time were a natural extension of our daily experiences. Relativity theory on the other hand requires something more abstract, but also, many believe, more elegant and revealing. In 1908 Hermann Minkowski pointed out that peculiar effects like length contraction and time dilation don't appear quite so unnatural if we cease thinking about space *and* time altogether, and think instead about *spacetime*. This is not just a four dimensional monstrosity invented by mathematicians to confuse people, but a much more accurate and indeed simpler model of the real world than Newton's. Its significance is revealed by simple examples such as the spacetime extension of the human body. It obviously has extension in space (about six feet) and duration in time (about seventy years), so it therefore has extension in spacetime. What makes this statement more than a truism is that the two extensions, spatial and temporal, are not independent. This is not supposed to mean that tall people live longer, or anything of that sort, but that viewed from a rocket on Earth a man might look three feet tall and live for one hundred and forty years. A neat way of looking at this is to think of his physical length and the duration of his life as being merely *projections* onto space and time respectively of his more fundamental spacetime extension. As always with a projection, the extension of the image depends on the angle to the object, and this remains true in spacetime as well as space. It turns out that a change of speed acts rather like a rotation in spacetime;

specifically, by altering one's velocity, one rotates one's four dimensional body away from space into time or vice versa. Thus, the Earthman's spacetime extension remains unaltered when viewed from the rocket, he merely has three feet of his bodily length twisted into seventy years of his life!

Putting in some numbers reveals that a small length of time is worth an awful lot of distance. Not surprisingly, considering its basic role in the theory, the speed of light acts as the conversion factor. Therefore, one year of time corresponds to one light year (over six million million miles) of space; one foot works out at about a nonosecond (one billionth of a second).

There is more to spacetime than a convenient way of visualizing time dilation and length contraction. To the relativist, the world *is* spacetime, and one no longer thinks of objects moving about *with* time, but extended *in* spacetime. To help make this distinction more concrete, Figure 5 depicts a typical region of spacetime. Because four dimensions cannot be drawn on a sheet of paper, only two dimensions of space are shown; time runs vertically upwards, space horizontally.

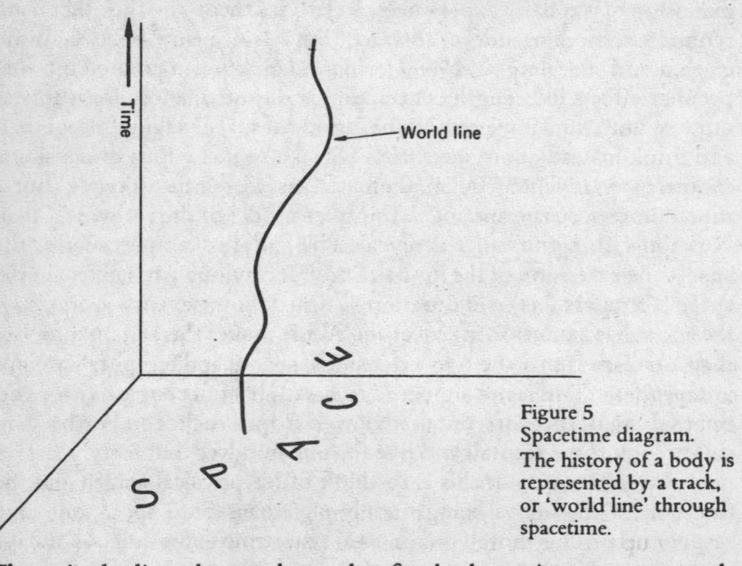

Figure 5
Spacetime diagram.
The history of a body is represented by a track, or 'world line' through spacetime.

The wiggly line shows the path of a body as it moves around. To avoid crowding the diagram, the spatial extension of the body is reduced so that it is represented by a line rather than a tube.

If the body remained at rest the line would be straight and vertical. When it accelerates, the line curves. The particle shown in Figure 5 first moves briefly to the right and back, then farther to the right, where it slows down and reverses again. These tracks in spacetime are called world lines, and they represent the complete history of the system of objects. If the diagram were enlarged to encompass the whole of spacetime (all the universe for all eternity) it would be a picture of the totality of events, and contain everything that physics can ever say about the world. Returning to the vexed question of what the world really is, we see that to a relativist it is spacetime and world lines. According to this picture of the universe the past and future are every bit as real as the present – indeed, no universal division into past, present and future can be made. It follows that things don't *happen* in spacetime, they simply *are*.

How are we to reconcile the static, once-for-all character of the relativist's universe with the world of our experience, where events occur, things change and our environment evolves. We do not perceive the world as a slab of spacetime threaded with lines, so what is missing?

Our actual experience of time seems to differ in two essential ways from the model of time as conceived in this theory. The first is the apparent existence of a 'now' or present moment. The second is the flow or movement of time from past to future. Let us begin by examining what is meant by the 'now'. The present plays two roles; it divides the past from the future, and it provides the leading edge of our consciousness as it cuts through time from past to future. Like the bow of a ship, the present trails behind it a wake of remembered events and experiences, but ahead lie unknown waters. These observations seem so natural that they must be above suspicion, yet a closer examination reveals several flaws. There cannot, of course, be 'the' present, because every moment in time is a present moment 'when it happens'. That is to say, there are past nows, future nows and now. But without any external quality against which to gauge it, there can be little that one can say about 'presentness' that is not tautologous.

A popular analogy is to regard the observer as a world line in spacetime, endowed with a little light. The light moves slowly and steadily up the line as the observer becomes conscious of successively later moments. However, this device is really a fraud, because it makes use of the idea of motion through time, and as such intuitively involves another time, external to spacetime, against which to gauge its progress. All this seems to imply that 'now' is just another way of labelling instants, and that there are as many nows as there are instants. We have already seen that 'now' is not, in any case, a universal characterization and that different observers will disagree about which events are simultaneous, but it appears that even for a particular observer the notion of *the* present doesn't really seem to make much sense.

A similar quagmire of contradictions and tautologies accompanies an examination of the idea of the flow of time. We have a deep psychological impression that time is moving from past to future, a development that sweeps the past out of existence and brings the future into being. Many examples can be found in literature to describe the impression: the river of time, time running out, time flying, time coming, time gone by, time waiting for no man . . . Saint Augustine regarded it thus

> Time is like a river made up of events which happen, and its current is strong; no sooner does anything appear than it is swept away.

So powerful are these kinetic sensations, that time *as* activity seems closest to our most fundamental experience. Yet where is the river in our space-time diagram? If time is flowing, how fast does it move? One second per second – one day per day: the question is devoid of meaning. When we observe an object moving through space we use time to gauge its rate of progress, but what can one use to measure the rate of progress through time itself? It may seem startling to ask: does time pass? Yet nothing that we can objectively measure in the world around us has proved that it does. There is no instrument which can record the flow of time, or measure its rate of passage. It is a common misconception that this is precisely the function of a clock. A clock, however, measures intervals of time, not the speed of time, the distinction being analogous to the difference between a ruler and a speedometer. The objective world *is* spacetime, with all events, for all times, included. There is no present, no past, no future.

One of the fascinations about time is the great disparity between our experience of it as conscious observers, and the objective properties it has in physics. We cannot escape the conclusion that the qualities of time we regard as the most vital – the divisions into past, present and future, and the forward movement of each division – are purely subjective. It is our own existence that endows time with life and motion. In a world devoid of conscious observers, the river of time would cease flowing. Sometimes the flow of time is referred to as merely an illusion brought about by deep-rooted confusion in the temporal structure of our language. Possibly an extraterrestrial intelligence would be utterly unable to comprehend the whole idea. On the other hand the confusion in our language (which undoubtedly exists) may well result from the aforementioned incompatibility between subjective and objective time. That is to say, it may be that our impression of a flowing time is not the result of muddled language and thinking, but vice versa: an attempt to use a vocabulary rooted in our basic psychological experience of time for the description of the objective physical world. Perhaps there 'really' are two types of time – psychological and objective – and we should develop two modes of description for talking about them.

I have put 'really' in quotation marks because the question of what is meant by 'real' is important here. Many people would argue that true reality must be independent of the conscious observer, so that subjective or psychological time, by its very personal nature, cannot qualify for the dignification 'real'. Nevertheless this personal experience is one apparently shared by all conscious observers who can communicate, so is perhaps just as real as hunger, lust or jealousy.

We must not assume that all vestige of past-future disappears from objective spacetime. Certainly one may define certain events to lie to the past or future of others, and verify this relation with laboratory instruments. Our spacetime diagram has a well-defined top (future) and bottom (past) asymmetrically related to one another, as a simple example will show. Figure 6 depicts a bomb which explodes into several fragments. It is a typical illustration of a time asymmetric change because it is irreversible: a motion picture film of the explosion, if played in reverse, would instantly be spotted as trickery because it would show the miraculous self-organization of the fragments into a well-ordered system. Similarly, turning diagram (i) upside down produces the same impossible sequence. The world is full of disordering influences like this that provide an objective, physical distinction between past and future. They do not, however, define *the* past, or *the* future. The distinction is the same as the asymmetry between left handed and right handed: the Earth rotates anticlockwise at the north pole, so it is always turning left, as it were, supplying a real distinction between left and right. Nevertheless we know it is absurd to ask which part of the Earth is farthest to the left, or

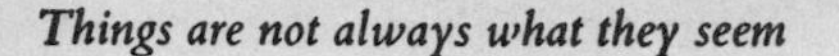

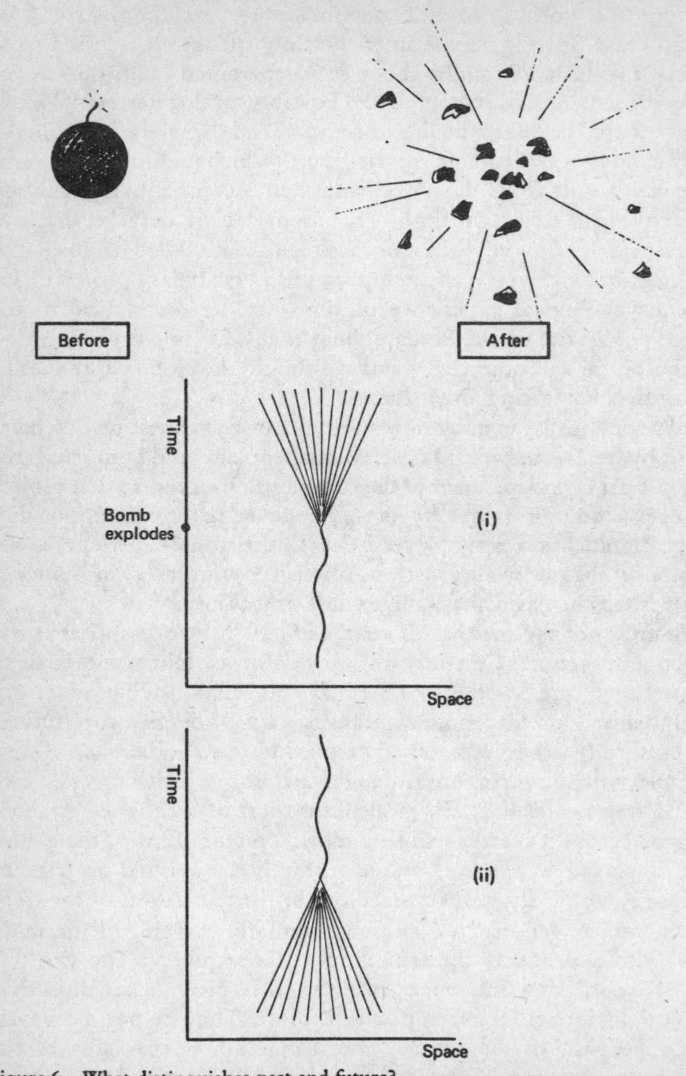

Figure 6 What distinguishes past and future?
Disordering defines an arrow of time, as the bomb fragmentation illustrates. In the spacetime diagram of this process (i) the past-future asymmetry is manifested by the branching world lines. The inverted picture (ii), representing the spontaneous assembly of fragments into a bomb, would be regarded as miraculous.

which country lies midway between left and right. Left and right define directions, not places. Similarly, past and future define temporal directions, not moments. Directions in, or through, time are objectively meaningful, but labelling events as past or future is apparently not. In chapter 10 the nature of time and our observation of it will be examined in greater detail.

The contrast between physical time and our experience of it emphasizes the crucial role that the conscious observer plays in organizing our perceptions of the world. In the old Newtonian picture, the observer did not seem to play an important part: the clockwork mechanism carried on turning, completely oblivious of whether, or by whom, it was being observed. The relativist's picture is different. Relations between events such as past and future, simultaneity, length and interval, become functions of the person who perceives them, and cherished impressions such as the present and the passage of time fade away from the world 'out there' altogether, residing solely in our own consciousness. The division between what is real and what is subjective no longer appears to be precisely drawn, and one begins to have misgivings that the whole idea of a 'real world out there' may crumble away completely. Later chapters will show how quantum theory requires the incorporation of the observer into the physical world in a still more essential way.

Einstein's 1905 theory of relativity overturned a lot of preconceptions about space, time and motion, but it was only the beginning. In 1915 he published an extended theory – the so-called general theory of relativity – in which still more extraordinary possibilities were proposed. We have seen how space and time are not fixed, but are in a sense elastic; they can stretch and shrink depending on who is observing them. In spite of this, spacetime, the four-dimensional synthesis of space and time, was assumed to be rigid. In 1915 Einstein proposed that spacetime itself is elastic, so that it can be stretched, bent, twisted and buckled. Thus, rather than merely providing the stage on which material bodies act out their roles, spacetime is really one of the actors. Naturally we cannot readily visualize what curvature is like in four dimensions, but mathematically four-dimensional curvature is no more remarkable than a curved line (one dimension) or surface (two dimensions).

Like all proper theories of physics, general relativity does not just predict that spacetime can be distorted, it provides an explicit set of equations which tell us how and by how much. The source of spacetime curvature is matter and energy, and Einstein's so-called field equations enable one to compute how much curvature there is at a point in space inside and around a given distribution of matter and energy. As would be expected, the bending of spacetime has a profound effect on the world lines of matter threading through it. As spacetime bends, so the world lines bend with it, and the question arises as to what physical effects a body will experience as a result of this rearrangement of its world line. It was explained in connection with Figure 5 that curvature of a world line corresponds to an accelerated motion for the body represented by the line, so the effect of spacetime curvature is to alter the motions of bodies located in it. Normally we regard change in motion as caused by a force, so the curvature manifests itself as a force of some sort. As all bodies, irrespective of their mass or internal structure, will similarly suffer the distortion, this force will have the distinctive property of affecting all matter indiscriminately without regard to their nature. A physical force with precisely this characteristic is known to us all – gravity. As found by Galileo, and since confirmed to phenomenal accuracy, all objects accelerate equally fast under gravity, whatever their mass or constitution, which implies that gravity is more a property of the surrounding space than of the bodies that move through it. In the words of John Wheeler, an American physicist who has advanced the theory of relativity enormously, matter gets its 'moving orders' directly from space itself, so that rather than regarding gravity as a force, it should be viewed as geometry. Thus 'space tells matter how to move and matter tells space how to curve'. General relativity is therefore an explanation of gravity as a distortion in the geometry of spacetime.

A number of famous experiments have measured spacetime distortion in the solar system. The planet Mercury was long known to suffer a mysterious disturbance in its motion: crudely speaking, its orbit twists by forty-three seconds of arc per century. Though minute, a twist of this magnitude can readily be measured and a straightforward application of Newton's theory of gravity does not account for it. When Einstein's paper was published, he predicted small corrections to the Newtonian theory as a result of spacetime curvature, and they worked out at precisely forty-three seconds of arc per century. This was a great triumph, but more was to come. In 1919 the astronomer Sir Arthur Eddington tested the theory of curved spacetime by looking at stars in the direction of the sun during a total eclipse (the eclipse enables stars to be seen during daytime even when located in the sky close to the sun). He found, as predicted, a tiny but measureable distortion in their positions when viewed in the proximity of the sun compared to their positions when the sun is in another part of the sky. Thus, as the sun wanders through the zodiac it buckles up our image of the backdrop of starry space.

A final crucial test of the theory was most elegantly performed using the Earth's gravity. According to general relativity, time is stretched or shrunk by gravity in the same way as it is by rapid motion. Thus, clocks at the surface of the Earth should lose relative to clocks at a higher altitude where gravity is marginally lower. The effect is truly minute – a hundred-billionth per cent reduction in clock rate for each vertical kilometre – but such is the precision of modern technology that even this difference can be detected. In 1959 scientists at Harvard University used as a clock the natural internal vibrations of a nucleus of radioactive iron. A certain isotope of iron decays by emitting a gamma ray – a photon of light with an internal frequency of about three million million megacycles. The gamma rays were shot up a vertical tower 22½ metres tall, where they encountered more iron nuclei. Normally these nuclei would reabsorb the gamma rays, but because time 'runs more rapidly' at the top of the tower, the gammas find that the internal vibrations of the iron nuclei no longer match their own frequencies as they did at the base of the tower. Absorption is thus inhibited. In this way, the dilation of time by the Earth's gravity could be measured.

More recently, the time distortion due to the Earth's gravity has been checked by flying a hydrogen maser in a space rocket. Maser is an acronym for 'microwave amplification by the stimulated emission of radiation', and is a version of the laser which oscillates at short-wave radio frequencies in an extremely stable way. Using the maser's cycles as the ticks of a clock, scientists were able to monitor the time in the spacecraft relative to Earth, by comparison with grounded masers. At ten thousand kilometres up, time should be increased by about one half billionth compared to its rate at the Earth's surface. Though minute, this significant effect easily showed up on the masers, and the theory was confirmed. Time really does 'run' faster in space.

The time stretching effect becomes even more dramatic as gravity rises. On the surface of a neutron star (see page 36) the disparity between clock rates at the surface and a long distance away is as much as one per cent. Stars that are slightly more massive than a neutron star will be shrunk even more and their gravity will rise still higher. If a star with the mass equivalent of the sun were to shrink to a few miles in diameter, the time distortion around it would become enormous. The star itself would be unable to withstand its own weight and would violently implode, shrivelling away to nothing in a microsecond. Its gravity would become so intense that on a surface in space around the collapsed object, time would be literally slowed down to a halt relative to distant places. A distant observer would deduce that clocks at this surface are completely frozen in time. He could not actually see the clocks, however, as the escape of light from the surface is also frozen in time. The hole in space left by the retreating star is therefore black – a black hole. Black holes are thought by many astronomers to be a routine fate for stars a little more massive than our sun.

Of course, an observer falling into the black hole across this 'frozen surface' would not regard time as behaving in an unusual way. In his frame of reference events proceed with their usual regularity, so his time scale becomes more and more out of step with the distant universe. By the time he reaches the frozen surface, which may only appear to him a duration of a few microseconds, all of eternity will have passed by elsewhere, and the cosmos will have burned out. The temporal dislocation escalates without limit, so that when he finally enters the black hole region, he is beyond time as far as the external world is concerned, one implication being that he can never re-emerge from the black hole in our universe. To do so would involve travelling backwards in time, reappearing from the hole before he even fell inside.

Though it is beyond eternity, the inside of the black hole is a region of spacetime much like any other as far as its local properties are concerned. Naturally the intense gravity causes the falling observer to feel a bit uncomfortable as his feet try to fall at a different rate from his head, but the progress of time is quite normal. The question of the fate of the falling observer is a curious one. It is conceivable that he falls right through the hole and emerges in another universe entirely, though what little evidence we have suggests that this will not happen. If he cannot return to our universe, cannot reach another, and cannot prevent himself continuing to fall inwards, where does he go? In chapter 5 we shall see that he is forced to leave spacetime altogether, and cease to exist as far as the known physical world is concerned. Black holes will also have an important part to play in later chapters on the question of how special the universe is.

The introduction of gravity into relativity further undermines the concreteness of the world. Spacetime, instead of just an arena, now becomes dynamical, able to move, change, twist and turn. We can no longer adopt the Newtonian view of seeking to understand the development of the world *in* time, we must also take into account changes in the spacetime fabric itself. The price paid for having a mutable spacetime is that it may actually contrive to smash itself out of existence. Following some convoluted motion that is intimately interwoven with the condition of matter and energy, Einstein's equations predict that situations can occur (such as at the centre of a black hole) where spacetime focuses its curvature without limit. In the escalating gravity, the violent distortion of spacetime becomes ever greater until it essentially comes apart at the seams. Some astronomers believe this is what will happen eventually to the entire universe; a catastrophic kamikaze plunge to extinction.

Gravity is a cumulative force, so it is no surprise that its effects are most pronounced in the subject of cosmology – the large scale structure of the universe. There are two ways in which the elasticity of spacetime might be important. The first, originally proposed by Einstein himself, is that space might not be infinite in extent but, like the surface of the Earth, curve up 'round the other side' of the universe and join together to form a hypersphere – a higher dimensional version of a spherical surface. We cannot envisage a hypersphere mentally, but we can compute its properties, one of these being the ability to circumnavigate the cosmos by travelling always in the same direction until one returns to one's starting point from the other way. Another is that, though the volume of space is limited, there is nowhere a barrier or boundary, nor is there a centre or edge. (All of these properties are shared by a spherical surface). At the moment, it is not known if there is sufficient matter in the universe to cause this complete topological closure.

The second possibility for elastic spacetime is that, on a cosmological scale (i.e. over distances much larger than galaxies) space might not be static, but stretching or shrinking. In the late 1920s the American astronomer Edwin Hubble discovered that the universe is, in fact, expanding; that is, space is stretching everywhere in what appears to be a very uniform fashion, a fact of some significance to which we shall return later. Hubble noticed that distant galaxies seem to be receding from us and each other as they are stretched apart by the expanding space. The evidence for this comes from the shift in wavelength of light, already discussed on page 37 in connection with the binary pulsar. For visible light, a lengthening of light waves emanating from a distant galaxy makes its colour appear more red than if it were stationary relative to us. The cosmological redshift increases in direct proportion to the distance of a galaxy from us, which is just the pattern of change that would result if the expansion motion were uniform, and taking place throughout the universe. The fact that all the galaxies seem to be receding from us does not mean that we are located at the centre of the cosmos, for the same pattern of recession would be visible from any other galaxy. Thus, the galaxies are not expanding away from anywhere special; there is no centre or edge of the universe that we can discern, even with our largest telescopes.

If the galaxies are moving farther apart it follows that they must have been closer together in the past. Looking out at distant regions of the universe enables astronomers to see back in time, because light from the most distant objects currently visible in telescopes may take several billion years to reach us, such is their distance. Telescopes therefore give us an image of what the universe looked like billions of years ago. With radio telescopes, the look-back time can be increased to about fifteen billion years, when a remarkable feature is found. The galaxies cease to exist, and indeed, all the structure we now observe – stars, planets, even normal atoms – could not have been present. This early epoch will turn out to play a central role in the theme of this book, and will be thoroughly discussed in chapter 9. Now it need only be mentioned that the expansion of the universe was then much more rapid than today, and the contents enormously compressed and hot. This hot, dense, exploding phase has been called the big bang, and some astronomers believe it not only marks the beginning of the universe as we know it, but perhaps the beginning of time itself. The big bang was not, as far as we can see, the explosion of a lump of matter into a pre-existing void, for that would imply a central core and an edge to the distribution of matter. What the big bang really represents seems to be the edge of existence, a concept that will become clearer in the pages to come.

3. *Subatomic chaos*

Throughout history, man has regarded his relation to the world as two-fold: observer and participator. We are conscious of the physical processes which take place around us, constructing internal mental models which reflect this external activity. In addition we are motivated to act on the external world, in small ways by living our daily lives, and collectively in grand ways by using technology to modify our environment. In spite of its rather humble scope compared with the great cosmic forces, our technology is nevertheless a demonstration that the existence of the biological species called homo sapiens is playing a part in shaping the universe, albeit a rather small portion as yet. With the Newtonian revolution, the status of man as participator looked a little hollow. True, the participation can hardly be denied, yet in a mechanistic universe, mechanically motivated man cannot be distinguished from the machines of his technology, so from our efforts to transform the environment down to the tiniest wiggle of a finger, human actions appeared to be just as rigidly predetermined and mindless as the planetary motions.

Let us now examine the Newtonian view of man the observer. What is really meant by the act of observation? Newtonian mechanics conjures up the picture of a universe criss-crossed by a network of influences, every atom acting on every other with tiny, yet significant, forces. All the forces we know to exist share the property that they diminish with distance, which is why we do not worry about Jupiter's effect on the tides, or take into account the movement of the Andromeda nebula when flying a plane. If the forces did not fade out with distance, the affairs of the Earth would be dominated by the most distant matter, for there are many more far-flung galaxies than nearby ones. Nevertheless as far as Newtonian forces are concerned, some residual influence, however infinitesimal, will still act between particles of matter that are separated even by immense distances. This interweaving of all matter into a collective whole recalls Francis Thompson's words

> All things by immortal power,
> Near or far,
> Hiddenly,
> To each other linked are,
> That thou canst not stir a flower,
> Without troubling of a star.

It is clear that there is a philosophical problem concerning the contradiction between a universe integrated by invisible forces, and the procedure of determining the laws of nature by isolating a system from its surrounding environment, as discussed in chapter 1. If we cannot free matter from its web of forces, it can never be truly isolated, and the mathematical laws we deduce can be at best only idealized extrapolations from the real world. Moreover, the crucial concept of repeatability – that according to the laws, identical systems should behave in identical ways – is also denied. There are no identical systems. Because the universe changes from day to day and place to place, the cosmic network of forces can never be completely identical.

All these objections notwithstanding, practical science proceeds apace, on the basis that the influence of, say, Jupiter on the motion of a motor car is less than any instrument could conceivably measure. However, when it comes to making observations, it is precisely these minute forces which play the vital role. If it were not for the fact that *some* influence from Jupiter had a detectable effect we could never know of its existence. The inescapable conclusion is that all observation requires interaction, of some sort. When we see Jupiter, photons of sunlight reflected from atoms in the Jovian atmosphere traverse the several hundred million miles of intervening space, penetrate the Earth's atmosphere and impinge on cells in the retina where they dislodge electrons from the atoms therein. This merest brush of a disturbance sets up a tiny electric signal which, when amplified and propagated to the brain, delivers the sensation 'Jupiter'. It follows that, through this chain, our brain cells are linked by electromagnetic forces to the atmosphere of Jupiter. If the chain of interaction is extended by incorporating telescopes, our brains can couple to the surfaces of stars billions of light years away.

An important feature of all types of interaction is that if one system disturbs another, thereby registering its existence, then there will be an inevitable reaction back on the first system, which in turn disturbs it. The principle of action and reaction is familiar from routine measurements in daily life. To measure an electric current, an ammeter could be inserted in the circuit, the presence of which will partially impede the very current being measured. To measure the brightness of a light it is necessary to absorb some of the radiation as a sample. To measure the pressure of a gas we could let the gas operate a mechanical device like a barometer, but the work done will be paid for out of the internal energy of the gas, the state of which is thereby altered. If we want to measure the temperature of a hot liquid, a thermometer could be inserted in it, but the presence of the thermometer will cause heat to flow from the liquid to the thermometer to bring it up to a common temperature. The liquid will therefore cool somewhat so the temperature which we actually read will not be the original temperature of the liquid, but that of the disturbed system.

In all these examples, access to the condition of a physical system is obtained by the use of probes. Sometimes more passive techniques are available, such as when we measure the location of a body simply by looking at it, as in the case of Jupiter. However, in order to get any information at all, *some* sort of influence must pass from object to observer, though its reaction may be utterly negligible for practical purposes. In the case of Jupiter, it would be invisible if it were not for its illumination by sunlight. This same sunlight which, when reflected, stimulates our retinas, also reacts on Jupiter by exerting a tiny pressure on its surface. (Sunlight pressure leads to a noticeable and spectacular effect by producing the tails of comets.) Thus, we do not, strictly, see the 'real' Jupiter, but one disturbed by light pressure. Similar reasoning can be applied to all our observations of the world about us. We can never, even in principle, observe *things,* only the interaction between things. Nothing can be seen in isolation, for the very act of observation must involve coupling of some sort.

Observing Jupiter illustrates a situation where the observer has only partial control over the circumstances; the sun's light is provided free. Therefore, the light pressure reaction will occur whether we choose to look at the reflected sunlight or not. In this sense, it cannot be claimed that Jupiter suffers a perturbation *because* we choose to observe it, only that without such a perturbation we could never observe it. In the laboratory, as the above examples illustrate, the involvement of the observer and his equipment is more direct.

We now come to the crucial feature in the act of observation as envisaged in the Newtonian picture of the universe, a feature which was in due course overthrown with the inception of the quantum theory. First, if the laws of physics are known, then although the measurement or observation will necessarily involve a disturbance to the subject of scrutiny, this disturbance can be accurately computed, and allowed for when deducing the result. Thus, the measurement of a liquid temperature can be corrected by knowing the thermal properties of the thermometer and its initial temperature. In a world where every atomic motion is rigorously determined by mathematical laws one can, in principle at least, take account of even the minutest disturbance from the measurement process. Second, by sufficient ingenuity and technological skill it is possible, according to the Newtonian theory, to reduce the troublesome disturbances to an arbitrarily small amount. Newtonian mechanics provides no limit to how weakly two systems may interact. Therefore, if we wished to measure the location of a body without knocking it off course by light pressure, we could use a flash gun with arbitrarily short illumination time. True, it would be necessary to amplify the reflected light more and more strongly with each decrease in the total light delivered from the gun, but this is a matter of technology and money, not fundamental physics. The conclusion seems to be that, in principle at least, the inevitable disturbance of observation can approach as close as we like to the perfect limit of zero (though naturally it cannot reach it).

As long as science dealt with macroscopic objects, little attention was paid to the ultimate limits of measurability, for in practical experiments these limits were never approached. The situation changed about the turn of the century when the atomic theory of matter became well-established, and the first subatomic particles and radioactive emissions began to be investigated. Atoms are so delicate that forces which are, by everyday standards, incredibly minute, can nevertheless produce drastic disturbances. The problems of carrying out any sort of measurement on an object only ten billionths of a centimetre in size and weighing a millionth part of a billion billionth of a gram, without destroying, let alone upsetting it, are formidable. When it comes to studying subatomic particles such as electrons, one thousand times lighter and with no discernible size at all, profound problems of principle, as well as practical difficulties, arise.

As an introduction to the general concepts we could consider the problem of how to ascertain simply where a particular electron is located. Clearly it is necessary to send in some sort of probe to locate it, but how can this be done without disturbing it, or at least by disturbing it in a controlled and determinable way? A direct approach would be to try to see the electron using a powerful microscope, in which case the probe employed would be light. As in the case of Jupiter so incomparably more with an electron, the illumination will exert a disturbance due to pressure. If we send in a pulse of light, the particle will recoil. This is not a severe problem if we can calculate how fast and in which direction the recoiling electron departs, for then, knowing the momentary location, it is just a matter of computation to deduce where the recoiling particle is at later times.

To get a good image in the microscope it is necessary to have a large objective lens, otherwise the light, being a wave, cannot get through the aperture without distortion. The problem here is that the light waves bounce off the sides of the lens and interfere with the primary beam, the effect being to fuzz out the image and destroy some resolution. It is necessary to use an aperture very much greater than the size of the waves (i.e. the wavelength). This is the reason why radio telescopes must be made much larger than optical telescopes – because radio wavelengths are so long. It follows that to see an electron properly we must either use a wide microscope or very short wavelength light, otherwise its image will be too smeared for an accurate location measurement. In addition to this, it is a familiar experience at the seaside that when large ocean waves meet a post or pier, they part temporarily as they pass by the obstruction, but join up again round the back to proceed relatively undisturbed. Thus the wave shape carries very little information about the location or shape of the post. On the other hand, short-wave ripples are seriously disturbed by a post, and their wave pattern breaks up into a complicated shape. By observing the disrupted pattern, the presence of the post can be inferred. A similar situation occurs with light waves: to see an object, waves whose length is comparable with, or smaller than, the object of interest must be employed. To locate an electron, waves of the shortest possible wavelength should be used (e.g. gamma rays) because its size is indistinguishable from zero. In any case, it is not possible to fix its position to an accuracy better than one wavelength of the light employed.

At this stage the quantum nature of light plays a crucial part. In chapter 1 it was explained that light only comes in lumps or quanta, called photons, and that when an atom absorbs or emits light, it does so only in whole numbers of photons. This endows light with some of the qualities of particles, because the photons carry definite energy and momentum; indeed, light pressure can be regarded as nothing more than the recoil suffered by collisions with photons. It does not follow, however, that light actually consists of little localized corpuscles. A photon is not concentrated at a place, but spread out in a wave. The particle nature of the photon is only manifested through the way in which it interacts with matter. The energy and momentum carried by a photon go down in proportion to its wavelength, implying that radio-wave photons are exceedingly feeble entities whereas light, and especially gamma rays, pack a lot more punch. This presents us with a conundrum when trying to see an electron, because the requirement of using very short wavelength radiation to avoid a smudged image entails accepting a violent recoil from the high impetus of these energetic quanta. We are thus forced to trade accuracy of location for disruption in the electron's motion. The resulting dilemma is that to determine accurately the quantity of recoil we need to know the exact angle through which the photon rebounds, and this can only be achieved by using a very narrow aperture microscope (see Figure 7). But, as already discussed, this strategy will result in a smudged image and loss of information about the electron's position. Neither will using longer waves help to reduce the recoil, for then we are forced to use a still larger aperture microscope to avoid wave bunching, which inevitably introduces more uncertainty into the angle measurement.

It should be clear by now that the requirements for an accurate determination of both position and motion are mutually incompatible. There is a fundamental limitation to the amount of information that can be obtained about the state of the electron. Its location can be accurately measured at the expense of introducing a random and utterly indeterminable disruption in its motion. Alternatively, the motion can be kept under control, but a large uncertainty introduced into its position. This reciprocal indeterminism is not just a practical limitation connected with the properties of microscopes, but a basic feature of microscopic matter. There is no way, even in principle, to acquire precise information about both the position and momentum

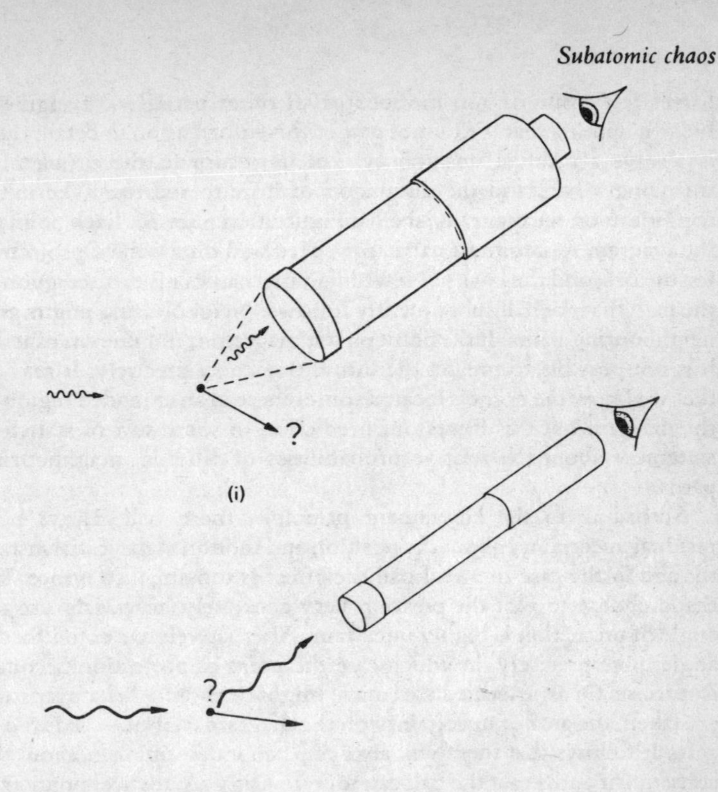

Figure 7 The uncertainty principle
(i) To locate the electron accurately demands a large microscope lens and short light waves, but the price paid is loss of knowledge about the recoil, for the rebounding photon could have entered the microscope anywhere within the cone of angles defined by the broken lines. Moreover, short-wave photons deliver a stronger blow to the electron. (ii) To compute the recoil accurately, one needs a narrow cone of angles and soft (long wave) photons, but this strategy yields a fuzzy image, thus destroying knowledge of the electron's position. Accurate knowledge of both position and motion is impossible, even in principle.

of a subatomic particle. These ideas are enshrined in the celebrated Heisenberg uncertainty principle, which relates the amounts of uncertainty in a mathematical formula, enabling the ultimate accuracy of any measurement to be assessed.

The implications of the uncertainty principle are iconoclastic. In chapter 1 we saw how knowledge of the position and motion of a particle was enough to determine its entire behaviour, if the impressed forces (or positions and motions of all other particles) are known. Now it appears that we cannot gather this information in detail; there is always a residual uncertainty. Let us return to the problem of throwing a ball, and the discussion of how to represent the initial conditions on the diagram labelled Figure 1 on page 26. Each point on the diagram represents a particular speed and direction of projection for the ball, and the laws of Newtonian mechanics give a prediction of the path that the ball subsequently follows. Neighbouring points give neighbouring paths. If the point on the diagram is not known exactly, it is not possible to predict the future trajectory precisely. It may be that we know the point is located somewhere in an extended region of the diagram but this limits our prediction to some sort of statistical statement about the relative probabilities of different neighbouring paths.

According to the Heisenberg principle, there will always be a residual uncertainty about the position and motion at the initial instant though in the case of a real ball the effect is too small to notice. We could choose to plot the position very accurately, in which case the angle of projection is highly uncertain. Alternatively we could fix the angle quite precisely, in which case the point of projection becomes imprecise. Or an intermediate course might be chosen. Whatever steps are taken, the area of uncertainty on the diagram cannot be reduced to zero. It follows that there will always be an indeterminism about the subsequent paths that the ball can follow. Only a statistical prediction can be made. On an everyday scale the quantum uncertainty is swamped by other sources of error, such as instrumental limitations, but the motion of 'atomic' balls is profoundly affected by quantum effects.

One's instinctive reaction to these ideas is to assume that the uncertainty is really a result of our lack of dexterity in atomic investigations, a consequence of our macroscopic size. It might be thought that the electron 'really' has a well-defined position and motion, but we are too ham-fisted to find out. This is generally believed to be quite wrong, for reasons which will be discussed at length in chapter 6. The uncertainty is apparently an inherent property of the microworld, not merely a consequence of our ineptitude in observing subatomic particles. It is not just that we cannot know what an electron is up to, it is that the electron simply does not possess a definite position and momentum simultaneously. It is an intrinsically uncertain entity.

It might be wondered whether anything at all can be said about the behaviour of such capricious and reticent objects. We cannot know the precise behaviour but only a collection of likely behaviours. The motion of an electron through space is thus not a well-defined affair but more like a spread of probability, with the available and possible paths spewing out after the fashion of a fluid. In 1924 Prince Louis de Broglie suggested that the behaviour of electrons was indeed analogous to that of a fluid; specifically he proposed that the possible paths spread out in the form of a wave. Thus, just as throwing a stone into a pond causes a pattern of ripples to emanate from a region, so too if electrons are liberated they will spill out in many directions, spreading about like the ripples on a pond.

There is far more to de Broglie's idea than a vague similarity of migration. The motion of a wave is very special both physically and mathematically. One vital characteristic is the ability of waves to interfere with each other. The phenomenon of wave interference is familiar in daily life, and also plays a fundamental part in the quantum description of matter and the consequences to be explored later. A good place to see waves interfering is on the surface of a pond. If two stones are simultaneously thrown into the pond close together, each sets up a sequence of radiating ripples. Where the two sets of ripples cross each other, a systematic pattern of peaks and troughs in the water surface is established. The pattern arises because where the wave peaks of one set coincide with those of the other, the disturbance is reinforced, but if a peak from one meets a trough from the other cancellation occurs and the water surface is relatively undisturbed.

In the 1920s physicists realized that if de Broglie was right, interference patterns ought to show up when beams of electrons are superimposed, for then the wave motions of each beam will overlap those of the other. Suddenly the experiments of Davisson, discussed in chapter 1, took on a new significance. Davisson found that electrons, when scattered from the surface of a nickel crystal, rebound in a whole succession of beams which subsequently overlap. In 1927 he proved beyond doubt that the overlapping beams enhanced or cancelled in a classic wave-interference pattern. The conclusion was startling: electrons behave like waves as well as particles.

What does this mean? Earlier we saw that light waves behave in some respects, though not all, like particles called photons. Now we seem to have a comparable duality of identity for electrons. However, it is most important to understand that the wave nature of electrons does not imply that an electron *is* a wave, only that it moves like one. Moreover, the wave being discussed is not a wave of any sort of substance or material, but a wave of probability. Where the wave disturbance is strongest, there the electron is most likely to be found. In this respect one is reminded of a crime wave which, when it spreads through a district, enhances the likelihood of a felony. It too is not a wave of any substance, but only one of probability.

These are challenging and provocative ideas, but they are also subtle and bewildering. A better understanding can be obtained by carefully studying a situation where both the wave and particle nature of electrons or photons are manifested in conjunction. One example is the so-called two-slit experiment. The layout is shown in Figure 8, and consists of an opaque screen containing two narrow nearby parallel slits. The slits are illuminated by a pinhole of light, and their images allowed to fall on another screen placed on the remote side. If one slit is temporarily blocked off, the image of the other will appear as a strip of light located in line with the open slit. Because the open slit is narrow, the light waves suffer distortion in squeezing through, so some light spills off the sides of the band giving it a fuzzy-edged

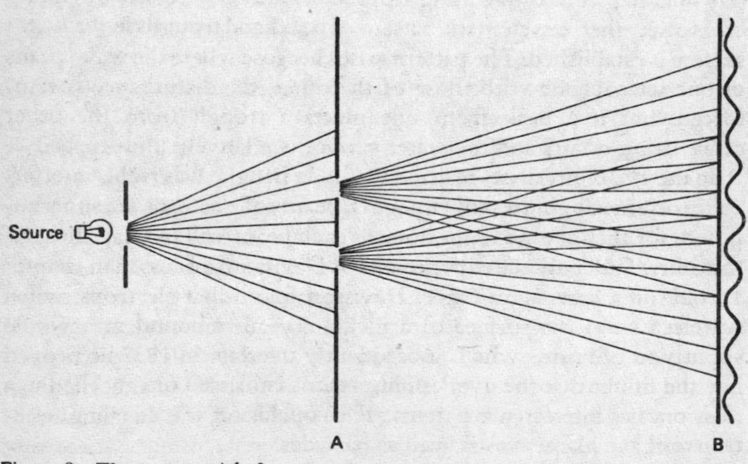

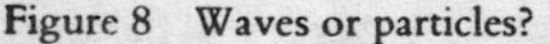

Figure 8 Waves or particles?
In this two-slit experiment electrons or photons from the source pass through two nearby apertures in screen A and travel on to strike screen B, where their rate of arrival is monitored. The curve graphed alongside B plots this rate. The pattern of peaks and troughs indicates a wave interference phenomenon.

appearance. If the slit is very narrow the light can be spread over quite a wide area. When the first slit is blocked and the other opened, a similar image is seen, but shifted slightly to line up with the other slit.

The surprise comes when both slits are open together. From the foregoing it might be imagined that the double-slit image would consist of two superimposed single-slit images, appearing as a double band of light overlapping somewhat due to the fuzzed out edges. In fact, what is seen is a sequence of regular stripes, consisting of dark and light fringes, first noted by the English physicist Thomas Young in 1803. The curious pattern is precisely the phenomenon of wave interference mentioned above. Where the light emanating from the two slits arrives out of step, i.e. the peaks of the waves from one slit meet the troughs from the other, the illumination is cancelled.

The experiment can be repeated with electrons rather than light, using a TV screen as a detector. Here we should remember that each individual electron is definitely a particle. Electrons can be counted one by one, and their structure probed using high energy machines. As far as we can tell they have no internal parts and no discernible extension. The slits are sprayed through a small hole by electrons from a type of gun. Those electrons that make it through one or other of the slits will proceed to the detector screen and strike it, delivering their energy in a little flash of light. (This is the principle of the TV image.) By monitoring the flashes, a record can be kept of exactly where the electrons arrive, and the way they are distributed across the detector screen can be determined.

Let us consider what happens when only one of the slits is open and the other temporarily blocked off. The stream of electrons spilling through the open slit will spray outwards and shoot across the intervening space to the detector screen. Most of them will arrive close to the region in line with the open slit, though some of them will spread out either side. The distribution resembles the illumination pattern which is obtained using light. A similar distribution, displaced slightly to one side, would be obtained if the other slit were left open and the first one blocked off. The crucial feature of the experiment is that, once again, when both slits are in operation, the distribution of electrons shows the regular pattern of interference fringes, indicating the wave nature of these subatomic particles.

This time there is an almost paradoxical quality about the result, for suppose that the intensity of the electron beam is gradually turned down until only one electron at a time passes through the apparatus. It is possible to record each individual electron's point of arrival at the screen using a photographic plate. After some time a stack of photographic plates is built up, each featuring a single spot of light where a particular electron flashed its presence. What can we now say about the way the electrons are distributed across the screen? This can be determined by looking through the lined-up stack of plates, where all the individual spots will be seen superimposed, forming a pattern. The astounding thing is that the pattern appears precisely as it did when vast numbers of electrons were spilling through together, and also precisely as it did for the light waves (though perhaps a little more speckled if we are sparse with the electrons). Evidently, the collection of separate and distinct events, involving only one electron at a time, still shows the phenomenon of interference. Moreover, if instead of repeating the experiment, electron by electron, a whole collection of laboratories try it out independently, and they each pick just one photograph at random, then the collection of all these separate, independent photographs also shows the interference pattern!

These results are so astonishing that it is hard to digest their significance. It is as though some magic influence was dictating events in different laboratories, or at different times in the same equipment, in conformity with some universal organizing principle. How does any individual electron know what the other electrons, maybe in other parts of the world, are going to do? What strange influence discourages the electrons from visiting the dark areas of the interference fringes and directs them towards the more populous areas? How is their preference controlled at the individual level? Magic?

The situation seems even more bizarre when we recall that the characteristic interference pattern arose in the first place because waves from one slit overlap those from the other. That is, interference is definitely a property of *both* slits. If one slit is blocked, the pattern disappears. Yet we know that any individual electron (being a tiny particle) can only pass through just one of the slits, so how does it know about the condition of the other one? In particular, how does it know whether the other is open or closed? It seems that the slit through which the electron does not pass (and which is, by atomic standards, an enormous distance away) has as much influence on the electron's subsequent behaviour as the one it actually passes through.

We are now beginning to glimpse something of the profoundly peculiar nature of the subatomic world. In chapter 1 it was mentioned that an electron is not constrained by deterministic laws to follow just one path, and later it was shown how Heisenberg's uncertainty principle prevents an electron from possessing a well defined trajectory. With the two-slit experiment this inherent indeterminacy can be seen in action, for we must conclude that *potential* electron paths thread through both slits in the screen, and that in some strange way those paths that are not followed still influence the behaviour of the actual path. Phrased differently, the alternative worlds that could have existed, but did not come to do so, still influence the world that does exist, like the fading grin of the Cheshire cat in Alice's tale.

It is now possible to comprehend why the waves associated with electrons are not waves of electrons, but probability waves. The interference that occurs in the two-slit system cannot be between many different electrons, or the pattern would disappear when only one electron at a time is used. It is an interference of probability. A single electron's location probability can probe both slits and interfere with itself. It is the propensity for an electron to visit a certain region of space that is being interfered with. So it is that an individual electron is more probably directed towards a bright fringe area than a dark fringe area. Because of the inherent uncertainty in the location and motion which gives rise to the wavelike behaviour, it is not possible to predict where any given electron will end up, but one can say something about a whole ensemble of them on the basis of simple statistics. It is just this statistical distribution which is subject to wave motion and interference effects and which must be taken into account in any calculation.

This demonstrates quite clearly how electrons avoid collapsing on to the nuclei of atoms. Their probability waves sit pulsating around the atom in a regular pattern. Only certain stationary patterns will occur, for if the wave disturbance does not fit together properly, with peaks and troughs in the right relation, it will start overlapping and interfering with itself and end up cancelling to nothing. If that happened there would be zero probability (no chance at all) of finding an electron. The phenomenon is similar to the standing wave pattern of air in a particular set of organ pipes – only certain well defined notes are allowed, because the air wave patterns must fit into the geometry of the pipes. So, too, only certain notes, i.e. frequencies or energies, are allowed round the atom. The characteristic colours emitted in transitions between these allowed energy levels is the visual evidence of this subatomic music. And just as there is a lowest note in an organ pipe, so there is a lowest energy level for an atom.

This is undoubtedly a great triumph for our understanding of the subatomic world, because the instability of atoms against collapse was one of the main mysteries which first motivated the rejection of Newton's physics for the atom. The facts that waves in musical instruments can produce a variety of discrete notes and that atoms can emit certain characteristic frequencies of light do not appear at first sight to be related, yet the wave nature of quantum matter reveals the beautiful unity of the physical world and shows that these phenomena are essentially the same. We can therefore regard the spectrum of light from an atom as similar to the pattern of sound from a musical instrument. Each instrument produces a characteristic sound, and just as the timbre of a violin differs markedly from that of a drum or clarinet, so the colour mixture of light from a hydrogen atom is characteristically distinct from the spectrum of a carbon or uranium atom. In both cases there is a deep association between the internal vibrations (oscillating membranes, undulating electron waves) and the external waves (sound, light).

Before leaving the two-slit experiment, an amusing feature must be described. Can an electron *really* know whether the other slit is open or closed? To find out we could attempt the following manoeuvre. Place a detector in front of the slits and spot which one the electron is headed for, then act quickly and block the other one off. If the electron gets wise to this manipulation, the interference pattern will not appear when we combine all the results from many such experiments. On the one hand, it is almost impossible to believe that an electron 'over there' can actually know our intentions and modify its motion accordingly; on the other hand, we know that if one slit is permanently blocked there is no interference pattern. Surely unblocking the hole when there are no electrons around cannot affect matters – can it? Either way nature appears to be playing games with us.

A simple procedure for carrying out this experiment is to shine a beam of light across the entrance to the slits and look for a tiny flash as the electron passes through. Naturally we must take into account the recoil of the electron as the light bounces off it, and we are reminded of the problems involved with the microscope discussed on page 60. To fix which slit the electron is approaching we must use light whose wavelength is short compared to the slit separation, or we will not obtain a sharp enough image to tell which slit is the nearer. However, short wavelength light will cause a relatively large disturbance to the motion of the electron of interest, and it turns out that the recoil imparted by light whose wavelength is short enough is so great that it completely destroys the interference pattern anyway. The unpredictable recoils scramble up the regular tracery of fringes. It seems that nature automatically prevents us from answering the crucial question: does the electron know whether the other slit is open or closed? Electron interference is a phenomenon that involves both open slits, yet any particular electron can only pass through one. We now see that the interference will only occur if we don't inquire too closely which slit the electron chooses. Both must be left open; either offers a potential path, though only one can be the actual path. Which one we can never know.

The modern theory of quantum mechanics amounts to much more than hand-waving arguments about the accuracy of measurements and the general ideas of wave motion. It is a precise mathematical theory capable of detailed predictions about the behaviour of subatomic systems. Important physical properties, such as Heisenberg's uncertainty principle, are built into the theory at a fundamental level, and emerge naturally from the formal mathematics. In particular, the Austrian physicist Erwin Schrödinger discovered in 1924 the actual mathematical equation which governs the progress of the enigmatic probability waves, and professional physicists today undertake practical calculations which reveal the internal structure and motion of atoms and molecules by attempting to solve this equation. For example, the energy levels of atoms, and hence the frequencies of light that they emit and absorb can be computed, along with the relative strengths of the different colours. These calculations enable hitherto mysterious spectra, such as those from distant astronomical objects, to be identified with known chemicals. This is especially important for those very distant objects such as quasars, because the light by which we see them has been enormously red shifted due to the expansion of the universe, and may represent radiation that would be invisible to us, in the ultra-violet region, had the shift not occurred. Calculations enable spectra at all frequencies to be predicted.

Other calculations can reveal the nature of interatomic forces, which help bind atoms together to form molecules. When two atoms approach, their matter waves start to overlap, and important interference effects occur which can cause the atoms to adhere in a chemical bond. When many atoms are packed together in a regular array, as in a crystal, the waves of all the electrons are forced into a coherent periodic motion which enables them to penetrate great thicknesses of material with little resistance. A study of these electronic waves gives information about how metals conduct electricity and heat. Detailed calculations using quantum theory have built up a picture of the structure of crystals and other solid materials such as semiconductors, as well as providing a basis for understanding liquids, gases, plasmas and superfluids. In the nuclear domain also, the application of mathematical calculations using quantum mechanics provides much information on the internal nuclear structure, nuclear reactions such as fission and fusion, and the interaction of nuclei with other subatomic particles.

The mathematics involved is not of the familiar type based on arithmetic, but involves abstract mathematical objects that obey very peculiar rules of combination and have properties quite different from those of ordinary numbers. Although the details of this mathematics require many years to learn, something of its flavour can be given using elementary ideas. As always in science, the mathematics is a model which must mimic the behaviour of the real world. In pre-quantum days the state of a physical system was represented by a collection of numbers. For example, the state of a body is defined by specifying its position, velocity, rate of spin, etc. at each instant. If these quantities are measured then the particular numbers may be ascertained. The way in which the numbers at one instant are related to those at other instants is provided by so-called differential equations.

In contrast, quantum theory forbids us to assign definite numbers to all the physical attributes of a body simultaneously: we cannot specify both the position and the momentum, for example. Moreover, there will not be a unique, well-defined trajectory, but many possible paths. The state of a system must reflect these uncertainties and ambiguities, and the act of measurement, which disturbs the quantum system in a fundamental way, cannot merely correspond to discovering the numerical values of the various quantities. One way of modelling the fact that a particle may exist in a quantum state which permits many possible behaviours – many different worlds – is using the idea of a vector. Vectors are familiar in daily life as directed quantities: velocity, force and spin are all examples of quantities that have both a magnitude (large, small, etc.) and a direction (towards the north, along the vertical, etc.). In contrast, quantities like mass, temperature, speed and energy all have magnitudes, but are not directed.

One important property of vectors concerns the way in which they must be added. Unlike numbers, we cannot simply add two vectors by adding their magnitudes, for account must also be taken of their directions. For example, if two forces are opposed, as in a tug of war, they may cancel to zero, even though their individual magnitudes are considerable. These considerations make the rules of combination for vectors more complicated than arithmetic, but also endow them with a richer structure.

Just as the addition of vectors can be performed in many ways depending on their directions, so a vector may be divided up in many ways into other vectors. For example, pushing a car is most efficient if you get right behind it, but it is still possible to move it, though less easily, by applying an oblique pressure. In fact, whatever angle the push, some of the force will act along the direction of motion, provided it is not precisely perpendicular to the car. Mathematicians say the vector has one component along the car and one perpendicular to it. Depending on the angle of push, so the parallel component enjoys a greater or lesser quantity of the total force than the perpendicular component. Thus the vector (the push) may be decomposed into two vectors; one parallel to the car, the other perpendicular, in a variety of proportions depending on the angle. For an angle almost parallel to the car, the parallel component gets most of the force and is much greater than the sideways force, so this is the most efficient pushing position (see also page 117).

The idea of a vector being decomposed into other vectors that are perpendicular to each other is used in a curious way in quantum theory. Each possible world, i.e. each potential behaviour or path of a particle, is treated as a vector; not a vector in ordinary space, but an abstract quantity in an abstract space. Every vector is perpendicular to every other vector, so that the worlds are all distinct and none of them has a component along any other. The number of vectors needed, and hence the number of dimensions in the space, depends on the choices of paths available. Recalling the paths-through-the-park analogy described on page 30, it might be necessary to use an infinity of potential worlds, just as there are an unlimited number of possible routes through the park. This requires an infinite-dimensional vector space: such a thing cannot be visualized, but makes good sense mathematically. Equipped with this vector space, the physicist can describe the state of a quantum system as a vector in the space which may point in general at any angle. If it lies along one of the vectors corresponding to a particular world, then an observation will show the system to definitely have the particular state corresponding to that world, but if it has an intermediate direction between two world vectors then, as with the force acting on the car, there will be components along both. The one with the greater component will be the most probable world, and the other a potential, but less likely alternative. Of course, if several alternatives exist, the vector may have components along all of them, and this can remain true even if they are infinite in number. The angle of the vector determines which are the favoured, i.e. more probable, alternatives (see Figure 9).

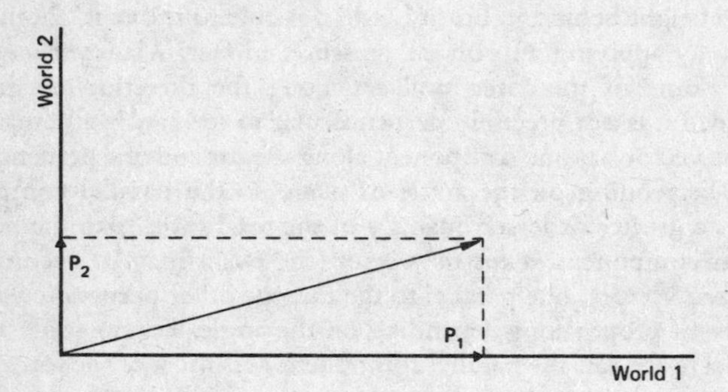

Figure 9 Superposition of worlds
The perpendicular arrows represent alternative worlds (e.g. electron goes through slit 1, or slit 2). The oblique arrow represents the quantum state which has a projection along both possibilities. As P_1 is longer than P_2 there is a greater probability that the world 1 is found to exist on measurement. If world 1 is indeed observed, the oblique vector suddenly and mysteriously collapses onto the horizontal arrow.

When it comes to making an observation, the system of interest, e.g. an atom, must obviously be found to reside in one particular state, e.g. the lowest energy level. This means that the original state, which may have been a superposition of many different alternative worlds, suddenly gets thrown or projected on to one particular alternative, a mysterious jump that will be examined in detail in chapter 7. In vector language this means that the act of observation causes the vector to rotate suddenly from some intermediate position in the abstract space to a new position where it lies exactly parallel to the vector which represents the world which is actually observed. This sudden jump in state, or rotation of the vector, reflects the fact that the observation inevitably disrupts the state of the system, as explained earlier in this chapter. Mathematically, therefore, measuring a quantity corresponds to suddenly rotating the vector in the abstract space.

Rotations provide another example of quantities that do not obey the rules of arithmetic. They too have magnitude (2°, 55°, a right angle, etc.) and a direction (clockwise, north-south, etc.) but adding rotations is even more complicated than adding vectors like forces, if their directions are different. In this case we must take into account not only the angle between the rotations, but the order in which we add them. When adding numbers, no account need be taken of the order of addition (e.g. 1+2=2+1), but this symmetry is not enjoyed by rotations. A simple example, which can easily be verified by the reader, is provided by rotating this book. Place it flat on a table in the usual reading position and tip it up through a right angle away from you, so it is upside down and vertical. Now rotate it clockwise through 90°. If the two above operations are performed in reverse order – a clockwise rotation followed by tipping away – the book will not end up in the same position. In fact, it will now be on its side instead of on its head. The example illustrates the general principle that rotations cannot be compounded by the usual rules of arithmetic, so they cannot be described by numbers for which the order of addition is immaterial.

These ideas fit naturally into the quantum scheme because the rotation of the state vector corresponds, as mentioned above, to a measurement, and the order in which two measurements are made will affect the end result. For example, if we measure the position of a particle, we destroy all knowledge about its motion. If, subsequently, we measure the motion, then the position becomes totally uncertain. When the measurements are carried out in reverse order – motion first, then position – we end up with a particle in a state with completely uncertain motion, which is not the same state as the end product from the former sequence. Thus, the order of observation, reflected in the order of rotation in the abstract vector space, is vital for the outcome. This feature is central to quantum theory, and appropriate mathematical objects, that do not obey the 1+2=2+1 rule of simple arithmetic, must be employed.

These powerful mathematical tools reveal new physics. Just as rotating a vector horizontally will affect its horizontal components but leave its vertical projection unaltered, so certain quantities are found to be 'perpendicular' to others and measurements can be performed on one without disturbing the other: for example, it may be possible to measure both the spin and energy of a particle simultaneously. Mathematical analysis uncovers which quantities are tied to others by the rotation-incompatibility property. These therefore satisfy uncertainty relations of the Heisenberg type. Important examples besides position and momentum are energy and time. It is not possible to measure a quantity of energy with complete precision unless an infinite quantity of time is available, a feature that will turn out to be of fundamental importance.

Most of the discussion in this chapter has been devoted to the curious wave-particle duality of the electron, but the considerations apply equally well to all microscopic matter. Since the Second World War hundreds of different types of subatomic particles have been discovered and all of them are subject to the rules of quantum mechanics. Indeed even whole atoms display the features of wave interference. There is no scale of size above which quantum matter turns into the Newtonian conception of 'ordinary' matter. Billiard balls, people, planets, stars – the entire universe is ultimately a collection of quantum mechanical systems, which implies that the old Newtonian picture of a clockwork universe moving in absolute determinism is false. In the everyday world quantum phenomena are far too small for us to notice; we do not see the wave properties of footballs, for example, because their wavelength is more than a billion billion times smaller than an atomic nucleus. Nevertheless the actual world is a quantum world, with all the far-reaching implications.

Lest we should feel that the mysterious matter waves are too far removed from daily experience to have any concrete significance, or that they may even be merely some outlandish construction of scientific imagination, we should be aware that they have now become a part of practical engineering. The electron microscope, a device capable of achieving enormous magnifications, bases its operation on electron waves as a substitute for light waves. By controlling the speed of an electron beam the wavelength can be manipulated, and it is a simple matter to obtain much shorter waves than visible light, enabling detail on a smaller scale to be resolved. Thus Davisson's curious patterns, so pregnant with implications for the nature of the universe, have a more prosaic, but more tangible impact on our lives.

4. *The strange worlds of the quantum*

We must now acknowledge that the microworld is governed not by deterministic laws that precisely regulate the behaviour of atoms and their constituents, but by randomness and indeterminacy. One aspect of this is that a particle such as an electron has wavelike behaviour, while electromagnetic waves can also display a particle aspect. There is no everyday counterpart of a 'wave-particle', so the microworld is not merely a Lilliputian version of the macroworld, it is something qualitatively different – almost paradoxically so. In this strange world of the quantum, intuition deserts us, and seemingly absurd or miraculous events can occur. In this chapter some of the consequences of the quantum theory will be explored, and the truly insubstantial nature of the apparently concrete world of matter will be described.

The Heisenberg uncertainty principle places restrictions on how accurately the location or motion of a particle is determined, but these two quantities are not the only ones that can be measured. For example, we might be more interested in the rate of spin of an atom, or its orientation. Alternatively, it may be necessary to measure its energy, or its lifetime before it changes state to a new energy. It is possible to analyze observations of these quantities in the same way that the gamma ray microscope, which was described in the previous chapter, was used to estimate the position and momentum uncertainty.

To illustrate these further possibilities, suppose we wish to determine the energy of a photon of light. According to Planck's original quantum hypothesis, the energy of a photon is related in direct proportion to the frequency of the light: double the frequency and the energy is doubled too. A practical means of measuring the energy is therefore to measure the frequency of the light wave, which can be achieved by counting the number of beats (i.e. peaks and troughs of the wave) that pass by in a given interval of time. For visible light this is very many – about a million billion every second. For the procedure to be successful it is obviously necessary for at least one, and preferably several, beats of the wave to occur, but each beat requires a definite interval of time. The wave must pass from crest to trough and back to crest again. A measurement of the light frequency in a duration less than this is clearly impossible, even in principle. For visible light, the duration involved is very short (a million-billionth of a second). Electromagnetic waves with a longer wavelength and lower frequency, such as radio waves, may require a few thousandths of a second to oscillate once. The photons of radio waves are correspondingly of very low energy. In contrast, gamma rays oscillate thousands of times faster than light, and their photon energies are thousands of times greater.

These simple considerations reveal that there is a fundamental limitation to the accuracy with which the frequency, hence energy, of a photon can be measured in a given interval of time. If the duration is shorter than one cycle of the wave, the energy is highly indeterminate, so there is an uncertainty relation linking energy and time that is identical to the position-momentum relation already discussed. To have accurate energy determination, a long measurement must be made, but if the time at which an event occurs is the quantity of interest, then its accurate determination can only be made at the expense of knowledge about the energy. There is thus a trade-off between information about energy and information about time similar to the mutual incompatibility of position and motion. This further uncertainty relation has the most dramatic consequences.

Before turning to the wider issues, one important point must be emphasized. The limitation on energy and time measurements, as with those of position and momentum, are not merely technological shortcomings, but absolute and inherent properties of nature. There is no sense in which a photon can be envisaged as 'really' possessing a well-defined energy at all times, even though we cannot measure it, or as being created at some particular instant with a definite frequency. Energy and time are incompatible characteristics for a photon, and which of the two is more accurately manifested depends entirely on the nature of the measurement that we choose to perform. We glimpse here for the first time the astonishing role that the observer himself will turn out to play in the structure of the microcosmos, for the attributes possessed by a photon appear to depend on just what quantities an experimenter may decide to measure. Moreover, the energy-time uncertainty relation, like the position-momentum one, is not restricted to photons, but applies to all subatomic activity.

One immediately observable consequence of energy-time uncertainty concerns the quality of light emitted by atoms. As mentioned on page 67, the pattern of colours radiated by different substances is decided by the spacing of the atomic energy levels, and this enables physicists to identify different chemicals simply by looking at the spectrum of their light. A typical spectrum, e.g. from a gas-filled fluorescent tube, will show a sequence of sharp lines representing the distinct frequencies (i.e. energies) of the light emanating from atoms of that type. Each line is caused by photons of one particular energy that are emitted when electrons in the gas atoms jump from higher to lower levels.

There is an important detail about these lines that beautifully illustrates the energy-time uncertainty relation. The emission of an individual photon occurs when a particular electron is knocked (e.g. by an electric current) into a higher energy level so that the atom is temporarily in an excited state. The excited state is only partially stable, however, and the electron soon drops back to its more comfortable low-energy state. The lifetime of the excited state depends on several factors such as the distribution of other electrons and the energy difference between the states, and can vary enormously from a billion billionth of a second up to thousandths of a second or even longer. If the lifetime is very short, then the energy-time uncertainty relation requires that the energy of the emitted photon will not be very well determined. From an observational point of view this means that a collection of identical excited atoms will not, when they jump back to their ground states, produce identical photons. Instead, the collection of photons will vary somewhat in their energy and hence frequency. Looking at the light from millions of atoms, the experimenter does not see a precisely defined colour, but a spread of colour peaked around the centre of the spectral line. The lines themselves, therefore, are not completely sharp, but have fuzzy edges, and the width of the line is directly related to the lifetime of the excited atomic state. Thus, a short-lived state gives a very broad line because its photons have a very uncertain energy, whilst a narrow line indicates a long lifetime and fairly sharply defined energy. By measuring the width of a line, physicists can infer the lifetime of the corresponding excited state.

One of the more remarkable consequences of the energy-time uncertainty relation is the violation of the most cherished law of classical physics. In the old Newtonian theory of matter, energy is rigorously conserved. There is no way that energy may be created or destroyed, though it may be converted from one form into another. For example, an electric fire converts electric energy into heat and light; a steam engine converts chemical energy into mechanical energy, and so on. However many times it gets converted, or divided up, the same total quantity of energy remains. This fundamental law of physics thwarted all attempts to invent a *perpetuum mobile* – a machine that could run without fuel – for it is impossible to get energy out of nowhere.

In the quantum domain, the law of conservation of energy seems to be under assault. To assert that energy is conserved obliges us, at least in principle, to be able to accurately measure the energy from one moment to the next to check that the total quantity has remained unchanged. However, the energy-time uncertainty relation requires that the two moments at which the energy check is carried out should not be too close together, or there will be an indeterminacy in the quantity of energy involved. This opens up the possibility that for very short durations the law of energy conservation might be suspended. For example, energy might spontaneously appear in the universe, so long as it disappears again during the time allowed by the uncertainty relation. Picturesquely speaking, a system may 'borrow' energy under a rather special arrangement: it must be paid back again very promptly. The greater the loan, the more rapid must repayment be. In spite of the limited term of the loan, we shall see that a great deal of spectacular work may be done with the borrowed energy while it lasts.

Because we are dealing with subatomic systems, the quantities of energy involved here are very small by everyday standards. There is no question, for example, of running a machine on borrowed energy, as the medieval inventors once hoped. The energy output from an electric light emitted in one second can only be borrowed via the uncertainty principle for a billion-billion-billion-billionth of a second. Put another way, the quantum loan mechanism can only enhance the output from an electric light by one part in one followed by thirty-six zeros.

In the subatomic domain things are different because the energies are so much smaller than in daily experience, and the activity is so vigorous that even durations of time that are utterly minute for us can permit a great deal to happen. For example, the energy required to raise an electron to an excited atomic state is so small that it can be borrowed for up to several million-billionths of a second. This might not sound very long, but it can lead to important effects. If a photon encounters an atom it can be absorbed, causing the atom to become excited as an electron is raised to a higher energy level. If the photon does not have quite enough energy to raise the electron the deficit can be borrowed enabling the excitation to take place temporarily. If the energy deficit is not too great, the loan could be quite long, perhaps a million-billionth of a second. This is long enough for the electron to circle right around the atom, and might be comparable with the lifetime of the excited state anyway. The result is that when the loan is paid back and the photon re-emitted, the atom has been excited long enough for its shape to be rearranged somewhat, so the emitted photon will not be ejected in the same direction as the original. This can be described by saying that the incoming photon is scattered by the atom, effectively bouncing off in a new direction.

The closer the photon is to the correct energy necessary to elevate the electron to an excited state, the less the loan, the longer the lifetime and the greater the scattering effect. Because energy is proportional to frequency, which is in turn a measure of the colour quality of light, it follows that different colours will scatter by different amounts. Thus, some materials are transparent to some colours and not others, so they appear as coloured when looked through. Preferential scattering of high frequency light also explains why the sky is blue: the white light from the sun contains many frequencies mixed together. The high frequencies correspond to colours such as blue and violet, the low frequencies to yellow and red. When the sunlight hits the atoms of air in the high atmosphere, some blue light is scattered out to colour the sky and the remaining light, robbed of its blueness, is rich in the lower frequencies, so appears yellow. That is why the sun has a yellow colour. When seen near the horizon, the greater depth of air through which the light passes enhances this effect, further depleting the shorter wavelengths, and the sun takes on a red colour.

As another illustration of energy uncertainty, consider the problem of rolling a ball over a hill. A low-energy roll will send the ball part of the way up the hill, where it will come to rest and then roll back. On the other hand, a high-energy roll will succeed in getting the ball to the top of the hill, where it will start to roll down the opposite side. The question then arises as to whether the ball can borrow enough energy through the Heisenberg loan mechanism to surmount the hill even though it has been projected at a very low speed. As a test of these ideas one can study the behaviour of electrons, which play the role of the balls, as they enter a field of electric force, which acts like a hill by slowing the electrons up. If electrons are shot at this electric barrier it is indeed found that some of them get through the barrier even when their energy of projection is well below that needed to overcome the obstacle on the basis of non-quantum considerations. If the barrier is thin and not too 'high' then the required energy can be borrowed for the short duration necessary for the electron to travel across the narrow gap. The electron therefore appears on the other side of the barrier, having apparently tunneled its way through. This so-called tunnel effect, like all phenomena governed by quantum theory, is statistical in nature: electrons will get through the barrier with a certain probability. The greater the energy deficit, the more improbable it is that the uncertainty principle will bale them out. For a real ball weighing one hundred grams, and a hill ten metres high and ten metres thick, the probability of the ball tunnelling its way through the hill while still one metre from the top is only one chance in the number one followed by a billion billion billion billion zeros.

Although irrelevant in macroscopic objects, the tunnel effect is vital to some subatomic processes. One of these is radioactivity. The nucleus of an atom is surrounded by a barrier similar to a hill, caused by the competition between electric repulsion and nuclear attraction. A constituent particle in the nucleus, such as a proton, is repelled strongly by the electric charges of all its neighbouring protons, but will not usually be ejected from the nucleus because this electric force is overwhelmed by the stronger attractive forces which bind the nucleus together. The latter, however, are very short ranged, and disappear completely beyond the nuclear surface. It follows that if a proton were to be plucked a short distance from the nucleus and released, it would be propelled away at high speed by the electric field, the nuclear force being unable to restrain it because of its isolation from the nucleus.

High speed emanations from radioactive atomic nuclei were discovered by Henri Becquerel in 1898 and called alpha rays. It was soon discovered that they are not rays at all, but particles; in fact they are composite bodies consisting of two protons bound together with two neutrons. An explanation for the escape of the alpha particles from radioactive nuclei involves the tunnel effect. The alpha particle, when inside the nucleus, does not have enough energy to overcome the bonds of nuclear force which glue it to the other particles. It remains trapped in the nucleus by a barrier of force which cannot be surmounted. Nevertheless by borrowing energy for a mere million-billion-billionth of a second - which is all it takes for the alpha particle to travel the ten-million-millionth of a centimetre through the nuclear surface – the particle can escape. For a loan of such short duration, the energy borrowed is comparable to the alpha-particle's existing energy, so that its behaviour is profoundly modified. It tunnels through the barrier and appears on the other side, where the now unchallenged electric force ejects it at enormous speed to become an alpha ray. In any given nucleus where this may happen, there is a definite probability that after a given time, alpha emission will have occurred. Thus, in a large collection of radioactive atoms, after twice this time, twice as many emissions will have occurred. Each radioactive substance therefore has a definite half-life against decay, the length of which depends sensitively on the size and thickness of the nuclear force barrier.

Behaviour that is just as remarkable can occur with particles whose energy exceeds that required to surmount a barrier. Due to the wave nature of matter, some waves are reflected back from the barrier, however energetic the particles may be. This implies that there is a definite probability of a particle rebounding from a barrier that is quite insubstantial. Indeed there is a chance, though an incredibly minute one, that a bullet will bounce back off a sheet of tissue paper.

In the early 1930s quantum theory was combined with special relativity, largely through the work of Paul Dirac, and it immediately brought a crop of new possibilities. Until then, the equation that physicists had been using to describe matter waves, presented by Schrödinger, was mathematically inconsistent with the principles of special relativity. Dirac sought a replacement equation, but found that a satisfactory form could not be achieved using the sorts of mathematical objects then known. It was necessary for him to invent a new type of quantity, called a spinor, which enabled his equation to possess the additional symmetries inherent in the theory of relativity. In many ways Dirac's equation predicts results that are little different from the earlier, non-relativistic equation. But two new, and profoundly significant, features emerged.

The first concerns the behaviour of particles when they are rotated. The laws of quantum mechanics make definite predictions about how bodies moving in curved paths, such as a circular orbit, should behave. Dirac found that to uphold these laws it is necessary to suppose that the particle itself is in some sense spinning (hence the term spinor). The motion of an electron around an atom, for instance, resembles that of the Earth (which also spins on its own axis) going around the sun. The intrinsic spin of the electron has a puzzling feature, however, not displayed by the spinning Earth. Imagine a ball spinning clockwise about a vertical axis. If the ball is tipped end over end, it will spin anticlockwise about the same vertical axis. Rolling the ball on further to take it through a full 360°, brings it back to its original condition, spinning clockwise once more.

This description seems so obvious that one is inclined to take it for granted and assume that it applies to all spinning bodies, including electrons. The extraordinary thing is that electrons do not simply return to their former state when rolled around once. In fact, they need two successive complete revolutions to bring them back to the same condition again. It is as though electrons have a double view of the universe, a feature quite unparalleled in macroscopic bodies and totally mysterious in daily experience. The origin of the electron's double nature concerns the behaviour of its associated wave under rotations. It turns out that after a single revolution, the wave returns, crudely speaking, with peaks and troughs interchanged, and only a second rotation restores their original configuration. All this indicates that the internal spinning motion of a subatomic particle is really very different in character from the simple minded idea of a rotating sphere. Nevertheless intrinsic spin can be measured in the laboratory and was, in fact, already inferred to exist on the basis of certain curious double lines in atomic spectra, before Dirac produced his explanation. Not all subatomic particles possess this peculiar Dirac-type spin with its two-fold property. Some particles do not spin at all, while others carry two or four units of spin, and the double image does not occur. The familiar particles, however – electrons, protons and neutrons – which make up ordinary matter, are all Dirac particles with the peculiar spin.

Dirac's work produced another sensational result that is even more extraordinary than intrinsic spin. The full implications of Dirac's equation took several years to sink in, but early on, in 1931, Dirac began to concentrate on a simple, though peculiar, feature of his new mathematics. Like all physicists, Dirac regarded equations as things to be solved, and assumed that every solution gave a description of some actual physical situation. So, for example, if the equation were used to examine the motion of an electron orbiting a hydrogen nucleus, then each solution should correspond to one particular possible state of motion. As expected, Dirac's equation possesses an infinity of solutions, one for each energy level of the atom, and more still for the motions of energetic electrons moving unbound by the attraction of the hydrogen nucleus. What was disturbing, however, was the discovery of a whole set of additional solutions that have no obvious physical counterpart. Indeed they appear at first sight to be completely meaningless. For every solution of Dirac's equation describing an electron with a given energy, there is a sort of mirror solution describing another electron with an equal quantity of negative energy.

Energy, like money, was formerly regarded as a purely positive quantity. A body can possess energy if it moves, or is electrically charged or excited in a whole variety of ways. Possibly all the energy can be drained out of a body so that it has zero energy, but what does an energy less than zero mean? What would a body with negative energy look like and how would it behave? Dirac was initially very distrustful of these mirror solutions, the obvious implication being that they are unphysical quirks – just excess mathematical baggage – not descriptions of the real world. Nevertheless experience has shown that when a mathematical solution to a law of nature exists, some physical counterpart usually also exists. Dirac examined what would happen if these curious negative energy states really were possible states of matter. He realized that they presented a great paradox, because they apparently allow any ordinary (i.e. positive energy) electron to jump down into a negative energy state by emission of a photon. Thus, what is usually taken to be the lowest energy (or ground) state of, say, the hydrogen atom would no longer be the lowest state after all, and we should be returned to the classical problem of how atoms are prevented from collapsing. Moreover, there is no limit to how negatively large the Dirac states can be, so that all the matter in the universe threatens to descend into a bottomless pit amid an infinite shower of gamma rays.

To avoid this catastrophe, Dirac made a remarkable proposal. What if ordinary matter were prevented from infinite descent because all the negative energy states were already occupied by other particles? The reasoning behind this idea stemmed from an important discovery made by the German physicist Wolfgang Pauli in 1925. Pauli studied the properties of spinning particles, not in isolation, but collectively. The curious double-handed nature of intrinsic spin is intimately related to the way in which two or more such particles respond to the proximity of others. As a result of their wave properties, two electrons will probe each other's presence, quite apart from the electric force between them, because the peaks and troughs of one wave will overlap and interfere with the peaks and troughs of the other. A mathematical study of this effect shows that a type of repulsion exists that prevents more than one electron at a time from occupying the same state. Crudely speaking, two electrons cannot be squashed too closely together. It is as though each electron possesses a small unit of territory that it will not allow to be invaded by its fellows.

The Pauli exclusion principle, as the territorial property became known, leads to some important effects. It implies that densely packed electrons will have a quite extraordinary rigidity, as the exclusive tendency prevents them being squashed into the same space. One place where squashing of matter is most ferocious is at the centre of a star. The immense weight of stars causes their cores to shrink under gravity to enormous densities – perhaps as high as a billion kilograms per cubic centimetre. So long as they continue to burn, further shrinkage is resisted by the production of huge quantities of heat that builds up the internal pressure. Eventually though, the fuel is burned up and progressive shrinkage occurs until the electrons start to feel uncomfortably close to their neighbours. The Pauli principle then sets in and tries to support the star against continued crushing. In stars like the sun, it will take another five thousand million years for this state of affairs to come about, but when it does the consequences are dire. The properties of this ultra-crushed matter are controlled dominantly by the electrons acting collectively. One result of the exclusive property is that the stellar material behaves in an odd way in the presence of heat. Instead of the injection of heat causing the matter to expand and cool, the heat remains trapped, driving up the temperature. If this continues to the point where new reserves of fuel start to burn, the contained heat builds up suddenly like an overheating pressure cooker and the core of the star explodes, a paroxysm not violent enough to shatter it into fragments, but traumatic enough to alter its structure drastically from a large, cool, red star to a blue hot giant. Eventually, all the fuel burns out and a star like our sun will end its days shrunk to about the size of the Earth, supported against further collapse by the electrons.

Another place where the rigidity between electrons plays a vital role is the interior of an atom. A large atom may contain dozens of electrons in orbit about the nucleus, and at first sight it seems that they must all collapse down to the lowest available energy. If this happened, all the electrons would be jumbled together in close proximity in a chaotic fashion, and it is doubtful if stable chemical bonds could ever form. What is actually found to be the case is that the electrons stack up in tidy shells around one another, the outer shells being prevented from collapsing on to the inner ones by the Pauli exclusion principle. Without this effect, all heavy atoms would implode into a mess.

Returning to the problem of Dirac's negative energy states, the Pauli principle offers a resolution of the paradox. Just as electrons in an atom are prevented from dropping to the lowest energy levels if these are already filled by other electrons, so ordinary electrons would be prevented from dropping into the bottomless pit if the pit were already filled up with electrons. The idea is simple but has an obvious shortcoming. Where are all these negative energy electrons (and other particles) that are blocking the pit? Being bottomless, an infinite number of particles is necessary to fill it up. Dirac's answer seems at first to be a trick. He argued that this whole infinity of particles is invisible, so that what we normally regard as empty space is not really empty at all, but filled with an infinite sea of undetected negative energy matter.

In spite of its outlandishness, Dirac's idea has some real predictive power. Let us consider, for example, how one of these invisible inhabitants of space would respond to the presence of a photon. Just like an ordinary electron, a negative energy electron can absorb the photon and use its energy to jump up into a higher energy state, so long as there is a space available. If the photon's energy is great enough, it can lift the negative electron right out of the pit into an ordinary, positive energy state, where there is plenty of room. Such an event would be witnessed by us as the abrupt appearance of a new electron out of nowhere simultaneously with the disappearance of the photon. Because an electron with positive energy is observable, the transition from negative to positive energy means that the electron simply materializes out of empty space. But that is not all. It leaves behind a hole in the negative energy sea. If the presence of a negative energy electron is invisible, its absence (i.e. the hole) ought to be visible. The absence of a negative energy, negatively charged particle should appear to us as the presence of a positive energy, positively charged particle. Thus, accompanying the newly-created electron will be a sort of mirror particle with opposite, positive, electric charge.

Dirac's theory therefore predicts an entirely new type of matter, now known as antimatter. An energetic photon should be able to create an electron-antielectron, or proton-antiproton, pair. In 1932 an American physicist, Carl Anderson, discovered an antielectron (usually called a positron) amid the subatomic debris of a cosmic ray shower. Since then, hundreds of antimatter particles have been produced in the laboratory, in dramatic confirmation of Dirac's equation. As expected, antimatter does not survive for long. A gap amid the negative energy sea will be sought by any positive energy particles above it. If an ordinary electron meets such a hole, it will disappear into it and vanish from the universe, emitting a gamma ray in payment for its drop in energy. This process is the reverse of pair creation, and is seen as the encounter of an electron with a positron followed by their mutual annihilation. Thus, whenever antimatter meets matter, explosive disappearance results.

The idea of matter being created and annihilated is a consequence of the theory of relativity, which Dirac carefully embodied in his equation. We found in chapter 2 that if a body is accelerated to near the speed of light it will become progressively more ponderous in an attempt to prevent itself being pushed through the light barrier. The excess weight represents the conversion of energy into mass, that at lower speeds would go instead towards increasing the body's speed. It follows that mass is really just a form of locked-up energy. For example, a proton contains one million-billion-billionth of a gram of mass, but so concentrated is this pent-up energy that even such a minute amount of matter could produce a flash of light visible to the human eye up to ten metres away. The conversion of energy into matter explains the sudden appearance of particle-antiparticle pairs by the Dirac mechanism, the quantity of energy required being stipulated by Einstein's celebrated $E=mc^2$ relation. The reverse process, in which matter is converted into energy, also occurs in nuclear bombs and power plants, as well as in the sun, whose source of power is the disappearance of four million tons of mass each second.

If mass is just a form of energy, as Einstein claimed, then energy, like mass, ought to have weight. What happens to the four million tons of sun-stuff lost every second? The answer is that it gets converted into sunlight, so one second's worth of sunlight should weigh four million tons. How can this be tested? The total quantity of sunlight striking the Earth per second weighs a paltry two kilograms, so collecting sunlight and weighing it is futile. Surprisingly, a better strategy is to weigh the still feebler light from distant stars. By using the sun's gravity to enhance the light's weight somewhat over that of the Earth, a starbeam grazing the edge of the sun can be weighed by watching it sag in the sun's gravity. This is precisely what Eddington did during the 1919 solar eclipse, as mentioned on page 50.

Although impressive, Dirac's theory of an invisible negative energy sea of particles is hard to swallow literally. Later mathematical developments showed that this model is really only heuristic, and that the Dirac equation requires further mathematical processing before it can give a comprehensive account of the appearance and disappearance of matter. In the more modern theory, pair creation and annihilation occur as before, but the difficulties with the negative energy states do not arise in the same way.

When the possibility of particle pair creation is combined with the Heisenberg energy-time uncertainty relation, some dramatic new effects are possible. To lift an electron out of the negative energy sea and thereby create an electron-positron pair requires a gamma ray of energy at least equal to $2mc^2$, twice the right-hand side of Einstein's equation. However, this rather large quantity of energy can be borrowed for about a thousand-billion-billionth of a second, enabling an electron-positron pair to come fleetingly into existence before fading away again. These phantom pairs fill the whole of space. What we normally regard as empty space is in fact a sea of restless activity, full of all sorts of impermanent matter; electrons, protons, neutrons, photons, mesons, neutrinos, and many more species of matter, each existing only for the briefest fraction of time. To distinguish these temporary interlopers from the more permanent form of matter that we all know, physicists use the word 'virtual' for the former and 'real' for the latter.

This ghostly mêlée is not merely theoretician's imagery, for the seething fluctuations can produce measurable effects, even among familiar objects. For example, the gel-like quality of certain paints arises from intermolecular forces induced by these vacuum fluctuations. It is also possible to disturb the vacuum by introducing matter.

A metal plate, which reflects light, will also reflect the evanescent virtual photons in the vacuum. By trapping them between two parallel plates their energy can be altered slightly, leading to a measurable force on the plates.

These new possibilities drastically modify the physicist's image of a subatomic particle. An electron, for instance, can no longer be regarded as a simple point object, for it is continually emitting and reabsorbing virtual photons via the Heisenberg energy-loan mechanism. Each electron is therefore surrounded by a cloud of virtual photons, and if we probe more closely we deduce the presence also of virtual protons, mesons, neutrinos, and every other species of particle, buzzing around the electron in a beehive of activity. Every subatomic particle is really dressed up in this sort of elaborate and complex cloak of virtual matter.

Sometimes the virtual cloud leads to unexpected physical effects. For example, the neutron is an electrically neutral particle, as its name suggests, so it does not carry any overall electric charge. However, every neutron is dressed by a cloud of virtual particles, some of which are electrically charged. There will always be an equal number of positive and negative charges present, but they need not be at the same place. The possibility therefore arises that the neutron may be surrounded by shells of electrically charged virtual particles, such as mesons. Thus, when an electron is shot at a neutron it will scatter off this electricity, enabling the charge distribution around the neutron to be mapped. Furthermore, the neutron, being a Dirac particle, possesses intrinsic spin, which means that as it rotates it drags around these shells of charge, setting up miniscule electric currents. These currents produce a magnetic field which can be measured in the laboratory. When this was first done in 1933 it caused consternation among physicists, who did not expect an electrically neutral object to have a magnetic field.

We can envisage a particle transporting a whole retinue of virtual particles along with it. None of the virtual particles lives long enough to acquire the status of an independent entity, for it is soon reabsorbed by the parent. Each virtual particle in turn carries its own sub-cloud of other virtual particles whose existence is still more evanescent, and so on, ad infinitum. If for any reason the parent particle were to disappear, the virtual particles could not be reabsorbed, and would be promoted to real ones. This is what happens when matter meets antimatter; for example, when a proton meets an antiproton they suddenly disappear and some mesons, or perhaps photons, from the virtual cloud are left with nowhere to go. They therefore appear in the universe as new, real particles of matter, their Heisenberg loan having been cleared once and for all by the mass-energy of the sacrificed proton-antiproton pair.

Many other subatomic phenomena can be explained with the help of the energy-time uncertainty relation. One of the fundamental problems of microphysics is to explain how two particles can act on each other with an electric force. Before quantum theory, physicists envisaged each charged particle as surrounded by an electromagnetic field, which acts on other charged particles nearby to cause a force. When quantum theory showed that electromagnetic waves are confined into quanta, atttempts were made to describe all effects of the electromagnetic field in terms of photons. However, when two electrons repel each other, no visible photons need be involved, and an explanation had to await the concept of a virtual particle or quantum developed in the 1930s. The electric force of attraction and repulsion can now be understood as follows. An electron is surrounded by a cloud of virtual photons, each of which only lives fleetingly on borrowed energy before being reabsorbed by the electron. When another charged particle comes close, a new possibility arises. A virtual photon might be created by one particle, but absorbed by another. A mathematical analysis reveals that this exchange of a virtual photon does indeed produce a force between the particles with just the characteristics expected of an electromagnetic field.

Following the success of explaining electric (and magnetic) forces in terms of the exchange of virtual photons, the question arose as to whether the other forces of nature – gravity and the nuclear forces – could be similarly described. The quantization of gravity is an important topic that will be deferred until the next chapter. The problem of the origin of the nuclear forces was tackled in the mid-1930s. The strong nuclear force which binds together the constituents of the nucleus (protons and neutrons) is quite different in character from the electromagnetic force. First, it is several hundred times stronger, but more problematic is the way in which it varies with distance. The electric force between charged particles slowly diminishes as they are separated, according to the so-called inverse square law. In contrast the nuclear force does not change much with separation at short range, until the particles are about one ten-million-millionth of a centimetre apart, when it suddenly drops to zero. The abrupt fade out of the nuclear force at such short range is vital for the structure and stability of nuclei, but means that it cannot be explained by the exchange of quanta that are too similar to virtual photons.

The solution was found by the Japanese physicist Hideki Yukawa in 1935. He proposed that nuclear particles exchange virtual quanta of a new type of field – a nuclear field – but unlike virtual photons, Yukawa's quanta possess mass. The way in which the presence of mass leads to a short-range force can easily be understood from the energy-time uncertainty relation. According to Einstein's relation $E = mc^2$, mass is a form of energy, and as we have seen, creating mass uses up a lot of energy. To make a virtual Yukawa quantum, it is necessary for a lot more energy to be borrowed to supply the mass. By the terms of the Heisenberg mechanism, the duration of the loan must be correspondingly shorter, so the distance to which the Yukawa virtual particle can travel is severely limited. Yukawa produced a complete mathematical treatment and discovered that the force between two nuclear particles should indeed fall off rapidly outside a certain range. As expected, the range is related in a simple way to the mass of the virtual quantum, and by using the experimental fact that the force fades out at about one ten-million-millionth of a centimetre Yukawa could fix the mass of his quanta to be around three hundred electron masses.

At this stage an exciting new possibility arose. Just as virtual photons can be promoted to real ones by annihilating the electrons to which they are tied, so Yukawa's virtual particles might be brought into independent existence if the nuclear particles to which they are tied are annihilated. For example, if an antiproton meets a proton, then the abrupt mutual disappearance of this pair should result in a shower of new particles. Yukawa called them mesons, as their mass lies somewhere between the electron and the proton. About ten years later, Yukawa's mesons were discovered, like Dirac's positron, in the subatomic debris of cosmic rays. Nowadays, they are routinely produced, by the annihilation of antiprotons and many other processes, in giant accelerator machines.

Although many of the ideas discussed in this chapter have been rather elementary, and really require a full mathematical treatment to make them accurate and precise, nevertheless the results are far-reaching. The apparently concrete world around us is seen to be an illusion when we probe into the microscopic recesses of matter. There we encounter a shifting world of transmutations and fluctuations, in which material particles can lose their identities and even disappear altogether. Far from being a clockwork mechanism, the microcosm dissolves into an evanescent, chaotic sort of place in which the fundamental indeterminacy of observable attributes transcends many of the cherished principles of classical physics. The compulsion to seek an underlying lawfulness beneath this subatomic anarchy is strong but, as we shall see, apparently fruitless. We have to face the fact that the world is far less substantial and dependable than envisaged hitherto.

5. *Superspace*

In the domain of the quantum the apparently concrete world of experience dissolves away among the mêlée of subatomic transmutations. Chaos lies at the heart of matter; random changes, restrained only by probabilistic laws, endow the fabric of the universe with a roulette-like quality. But what of the arena in which this game of chance is being played – the spacetime background against which the insubstantial and undisciplined particles of matter perform their antics? In chapter 2 we saw how spacetime itself is not as absolute or unchanging as traditionally conceived. It too has dynamical qualities, causing it to curve and distort, to evolve and mutate. These changes in space and time occur both locally, in the vicinity of the Earth, and globally as the universe stretches with its expansion. Scientists have long recognized that the ideas of quantum theory should apply to the dynamics of spacetime as well as to matter, a concession which has the most extraordinary consequences.

One of the more exciting results of Einstein's theory of gravity – the so-called general theory of relativity – is the possibility of gravity waves. The force of gravity is in some respects like the force of electricity between charged particles, or the attraction between magnets, but with mass playing the role of charge. When electric charges are violently disturbed, such as in a radio transmitter, electromagnetic waves are generated. The reason for this can readily be visualized. If an electric charge is pictured as surrounded by a field, then when the charge is moved the field must also adjust itself to the new position. However, it cannot do this instantaneously: the theory of relativity forbids information to travel faster than light, so the outlying regions of the field do not know that the charge has moved until at least the light-travel time from the charge. It follows that the field becomes buckled, or distorted, because when the charge first moves the remote regions of the field do not change whereas the field in the proximity of the charge is quick to respond. The effect is to send a kink of electric and magnetic force travelling outward through the field at the speed of light. This electromagnetic radiation transports energy away from the charge into the surrounding space. If the charge is wobbled to and fro in a systematic way, the field distortion wobbles likewise, and the spreading kink takes on the features of a wave. Electromagnetic waves of this sort are experienced by us as visible light, radio waves, heat radiation, x-rays and so on, according to their wavelength.

In analogy to the production of electromagnetic waves we might expect the disturbance of massive bodies to set up kinks in the surrounding gravitational field, which will also spread outwards in the form of gravity waves. In this case, though, the ripples are kinks in space itself, because in Einstein's theory gravity is a manifestation of distorted spacetime. Gravity waves can therefore be visualized as undulations of space, radiating away from the source of disturbance.

When the nineteenth century British physicist James Clerk Maxwell first suggested, on the basis of a mathematical analysis of electricity and magnetism, that electromagnetic waves could be produced by accelerating electric charges, much effort was devoted to producing and detecting radio waves in the laboratory. The result of Maxwell's mathematics has been radio, TV and telecommunications in general. It might seem that gravity waves should prove equally important. Unfortunately, gravity is so weak that only waves carrying enormous energy will have any effect that can be detected using current technology. It is necessary for cataclysms of astronomical immensity to befall matter before detectable gravity waves are produced. For example, if the sun exploded or fell down a black hole, present instruments would easily register the gravitational disturbances, but even events as violent as supernovae explosions elsewhere in our galaxy are only just on the limits of detectability.

Gravity wave detectors, like radio receivers, operate on a very simple principle: as the space ripples wash through the laboratory, they set up vibrations in all the hardware. The ripples act to stretch and shrink space alternately in any particular direction, so all objects in their path also get stretched and squeezed by a minute amount, with the result that sympathetic oscillations can be induced in metal bars or incredibly pure crystals, of the right size and shape. These objects are very delicately suspended, and isolated from more mundane sources of disturbance, like seismic waves or motor cars. By looking for minute vibrations, physicists have attempted to detect the passage of gravitational radiation. The technology involved is mind-boggling: some bars consist of pure sapphire crystals as big as an arm, and the wobble-detectors are so sensitive that they can register a motion in the bar that is less than the size of an atomic nucleus.

In spite of this impressive instrumentation, gravity waves have not yet been detected on Earth to everyone's satisfaction. However, in 1974 astronomers discovered a peculiar type of astronomical object which gave them a unique opportunity to spot gravity waves in action. This object is the so-called binary pulsar, already mentioned in chapter 2 in connection with the speed of light. So accurately can astronomers monitor the radio pulses, that the most minute disturbances to the pulsar's orbit are detectable. Among such perturbations is a tiny effect due to the emission of gravity waves. As the two massive collapsed stars career about each other, intense gravitational disruption is created, with the result that a great deal of gravitational radiation is shaken off. The gravity waves themselves are still too weak to be detected, but their reaction on the binary system is measurable. Because the waves transport energy away from the system, the loss must be paid for out of the orbital energy of the two stars, causing their orbit to slowly decay, and it is this decay that astronomers have observed. The situation is rather like watching your electricity bill mount when your radio transmitter is switched on: the effect is not a direct detection of radio waves, but a secondary effect attributed to them.

The reason for this digression into the subject of gravity waves is that their cousins – the electromagnetic waves – were the starting point of quantum theory. As explained in chapter 1, Max Planck discovered that electromagnetic radiation can only be emitted or absorbed in discrete packets or quanta, called photons. We would therefore expect that gravity waves ought to behave similarly, and that discrete 'gravitons' should exist with properties similar to those of photons. Physicists support gravitons for stronger reasons than simple analogy with photons: all other known fields possess quanta, and if gravity were an exception it would be possible to violate the rules of quantum theory by letting these other systems interact with gravity.

Assuming gravitons exist, they will be subject to the usual uncertainties and indeterminacies that characterize all quantum systems. For example, it will only be possible to say that a graviton has been emitted or absorbed with a certain probability. The significance of this is that the presence of a graviton represents, crudely speaking, a little ripple of spacetime, so that uncertainty about the presence or absence of a graviton amounts to an uncertainty about the shape of space and duration of time. It follows that not only is matter subject to unpredictable fluctuations, but so is the very spacetime arena itself. Thus, spacetime is not just a forum for nature's game of chance, but is itself one of the players.

It may appear startling that the space we inhabit takes on the features of a quivering jelly, but we do not notice any quantum rumblings about us in daily life. Nor do sophisticated subatomic experiments reveal random and indeterminate jerks of the spacetime inside the atom; no sudden unpredictable gravitational forces have been detected. A mathematical analysis shows that none are expected: gravity is such a feeble force that only when huge concentrations of gravitational energy are present is spacetime distorted enough for us to notice. Remember that the entire mass of the sun will only distort the images of distant stars by a barely perceptible amount. On subatomic scales, temporary concentrations of mass-energy can be 'borrowed' through the Heisenberg uncertainty mechanism, so it is a simple matter to calculate the duration for a loan of enough mass-energy to put a really good bump into space. The Heisenberg principle requires that the greater the energy the shorter the loan, so because of the relative feebleness of gravity and the correspondingly intense packet of energy needed, only a very short loan indeed is possible. The answer works out to be the shortest time interval ever considered as physically relevant: sometimes known as a 'jiffy', there are no less than one followed by forty-three zeros (written 10^{43}) of them in one second, a duration so short that even light can travel a mere million-billion-billion-billionth of a centimetre in one jiffy - a full twenty powers of ten smaller than an atomic nucleus. Small wonder that we do not encounter quantum spacetime fluctuations in either daily or laboratory experience.

In spite of the fact that quantum spacetime inhabits a world within us more remote in smallness than the ends of the universe seem in their immensity, nevertheless the existence of the effects would lead to the most dramatic consequences. The commonsense picture of space and time is rather like that of a canvas on which the activity of the world is painted. Einstein showed that the canvas itself can move about and suffer distortions – spacetime comes alive. Quantum theory predicts that if we could examine the surface of the canvas with a super-microscope we should observe that it is not smooth, but has a texture or graininess caused by random and unpredictable quantum distortions in the spacetime fabric on an ultramicroscopic scale.

Down at the size of one jiffy still more spectacular structures appear. The distortions and bumps are so pronounced that they curl over and join up with each other, forming a network of 'bridges' and 'wormholes'. John Wheeler, the chief architect of this bizarre world of Jiffyland, describes the situation as similar to that of an aviator flying high above the ocean. At great altitudes he can only make out the gross features and sees the surface of the sea as flat and uniform, but on closer inspection he can make out the rolling swell that indicates some form of local disturbance: this is the large scale gravitational curvature of spacetime. Swooping down, he then notices the irregular small-scale disturbances – the ripples and waves superimposed on the swell: these are the local gravitation fields. Eventually, with the aid of a telescope he perceives that, on a very small scale indeed, these ripples become so distorted that they break up into foam. The apparently smooth unbroken surface is really a seething mass of tiny spume and bubbles: these are the wormholes and bridges of Jiffyland.

According to this description, space is not uniform and featureless but, down at these unbelievably small sizes and durations, a complex labyrinth of holes and tunnels, bubbles and webs, forming and collapsing in a restless ferment of activity. Before these ideas came along, a lot of scientists tacitly assumed that space and time were continuous down to any arbitrarily small scale. Quantum gravity suggests instead that our world canvas not only has texture, but a foam or sponge-like structure, indicating that intervals or durations cannot be infinitely subdivided.

Great mystification frequently surrounds the problem of what constitutes the 'holes' in the fabric. After all, space itself is supposed to be emptiness; how can there be a hole in something already empty? To answer this point it is helpful to visualize not Wheeler's wormholes, but holes in spacetime that are large enough to affect daily experience. Suppose there were a hole in space in the middle of Piccadilly Circus in central London. Any unsuspecting tourist would abruptly disappear on encountering this phenomenon, presumably never to re-emerge. We could not say what happened to him because our laws of nature are restricted to the universe, that is, to space and time, and say nothing of regions beyond their boundaries. Similarly, we could not predict what might come out of one – including what pattern of light. If nothing at all can emerge the hole would simply appear as a black blob.

There is no particular reason why our universe should or should not be infested with holes, or even complete edges. Figuratively speaking, God may have taken a pair of scissors to the spacetime canvas and lacerated it. While we have no evidence that this has happened on a Piccadilly-scale, something like it might be the case in Jiffyland.

A proper study of the branch of mathematics known as topology (the gross features and structure of space) reveals that holes in space need not cause the abrupt disappearance of objects from space. This may easily be seen by comparing space with a two-dimensional surface, or sheet, as we have done in our canvas and ocean metaphors. In Figure 10 two possibilities for holes in space are shown. In one a hole is cut in the middle of a roughly flat sheet: the sheet also has edges. The broken lines drawn on the sheet represent the paths of explorers who, like the imaginary ill-fated navigators of a previous century, vanish off the edge of the world (or into the hole). In the second example the sheet is curved over and rejoined with itself in the form of a doughnut, a shape known to mathematicians as a torus. The torus also has a hole in the middle, but its relation to the sheet is quite different from Figure 10(i). In particular, there is no abrupt edge, either bounding the hole or at the extremities, so explorers may crawl all over this surface without risk of leaving it: it is a closed, finite but edgeless space and is closer to the mathematician's view of the froth of Jiffyland.

It is entirely possible that the universe on a *large* scale has a shape analogous to the torus in Figure 10(ii) in which case space would not extend for ever, but curve back round on itself. Of course, it may not have a big hole in the middle – it could be more like a sphere – but in either case we could in principle travel all around it and visit every region. In colloquial jargon, we could 'do' the whole universe on a sort of cosmic package tour. And just as terrestrial globetrotters often leave London for Moscow but return from New York, so our intrepid cosmonauts might circumnavigate the cosmos, in what they regard as a fixed and straight flight path, returning from the direction opposite to their one of departure.

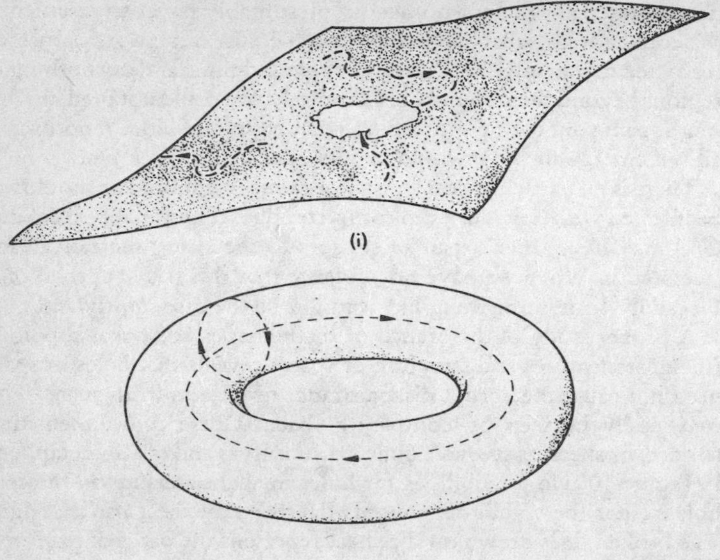

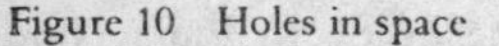

Figure 10 Holes in space
Space is here represented by a surface on which explorers crawl, leaving their tracks as broken lines. (i) The explorers 'fall off' the edge of the world or into the hole. (ii) They can circumnavigate the 'universe' without leaving the space – this surface does not have a boundary, even though it is limited in size and there is a hole in it.

The topology of the universe might be much more complicated than either the simple 'torus' or 'sphere', and contain a whole network of holes and bridges. One could imagine it as rather like a Swiss cheese, with the cheese being spacetime and the holes breaking it up into a complicated topology. In addition it must be remembered that the whole monstrosity is also in a state of expansion. Space and time would then be connected to themselves in a bewildering way. It would be possible, for instance, to go from one place to another by a variety of routes – each apparently a straight path – by threading through the labyrinth of bridges. The idea of a space bridge giving almost instantaneous access to some distant galaxy is much beloved of science fiction writers. The possibility of avoiding the long route through intergalactic space would be most appealing if giant wormholes really do thread the universe. Taking the canvas analogy, such a hole would be represented by curving the canvas over in a U shape and joining the

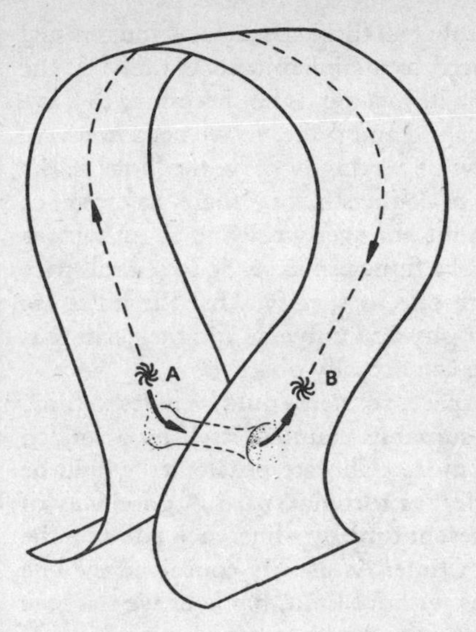

Figure 11 Space tunnel
Travelling from galaxy A to B through the tunnel saves the long route through intergalactic space (broken line).

two folds together at a certain place to form a tunnel (see Figure 11). Unfortunately, there is no evidence whatever that such features really exist, but neither can they be ruled out. In principle our telescopes should be able to reveal just what shape the universe is, but at present it is too difficult to untangle these geometrical effects from other, more mundane, distortions.

Still more bizarre possibilities come to mind. When our surface (i.e. space) is 'connected up' with itself, it could contain a twist, like the famous Möbius strip (see Figure 12). In this case it would no longer be possible to distinguish left-handed from right-handed. Indeed a cosmic circumnavigator might return as a mirror image of himself, with his left and right hands interchanged!

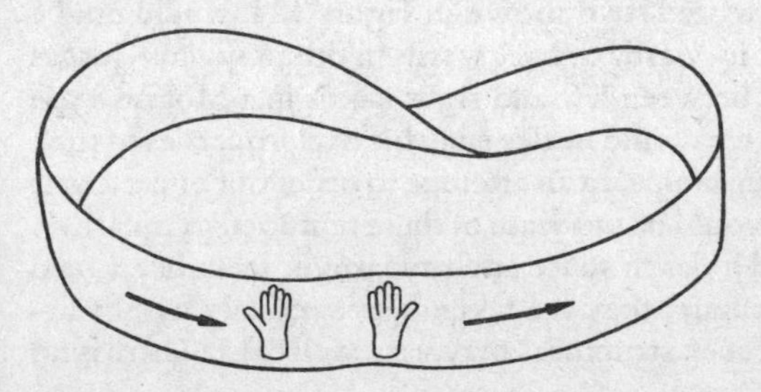

Figure 12
The Möbius strip has the strange property that a right-handed glove changes into a left-handed glove when transported once around the strip. (No distinction is to be made between the front and reverse surfaces of the strip.)

An important point to grasp is that all these spectacular and unusual features of space could be deduced by its inhabitants entirely on the basis of observations from within it. Just as it is not necessary to leave the Earth to conclude that it is round and finite, so we need not have the higher-dimensional overview of perceiving, say, the 'hole' in the middle of a 'doughnut' universe to deduce that it is there. Its existence has consequences for space without our ever worrying about what is 'in' the hole, or what is 'outside' the finite universe. So to regard space as full of holes does not require one to specify what the holes are physically – they are outside our physical universe and their nature is irrelevant to the physics that we can actually observe.

Just as there could be holes in space, so there could be holes in time. A crude cut in time would presumably manifest itself by a sudden cessation of the universe, but a more elaborate possibility would be closed time, analogous to spherical or toroidal space. A good way of visualizing closed time is to represent time by a line: each point on the line corresponds to a moment of time. As usually conceived the line stretches away in both directions without limit, but later we shall see that the line may have one, or two, ends: i.e. a beginning or end of time. However, the line could still be finite in length without having ends, for example by closing it into a circle. If time were really like this, it would be possible to say how many hours constituted the entire duration of time. Often closed time is described by saying that the universe is cyclic, with any event repeating itself ad infinitum, but this picture presupposes the dubious notion of a flow of time, sweeping us repeatedly round and round the circle. As there is no way to distinguish one trip around from the next, it is not really correct to describe such an arrangement as cyclic.

In a closed-time world the past would also be the future, opening up the prospect of causal anarchy and temporal paradoxes frequently discussed by science fiction writers. Worse still, if time joins up with itself similarly to the twisted strip shown in Figure 12 it would not be possible to distinguish forwards or backwards in time anyway – just as there is no distinction between left and right hands in a Möbius-type space. Whether or not we would notice such bizarre properties of time is not clear. Perhaps our brains, in an attempt to order our experiences in a meaningful way, would be unaware of these temporal gymnastics.

Although edges and holes in space and time might seem like a mad mathematician's nightmare, they are taken very seriously by physicists, who consider that such structures may very well exist. There is no evidence for the 'laceration' of spacetime but there is a strong suggestion that space or time might develop edges which have borders, or boundaries, so that rather than tumbling unsuspectingly off the edge of creation, we should be painfully and, it turns out, suicidally aware of our impending departure ('holes with teeth'). Glancing once again at Figure 10(i), it is clear that the hole which is simply cut in space starts abruptly. There are no warning features in the vicinity of the edge to herald the imminent discontinuity. Likewise with similar holes in time: nothing would herald the demise of the universe, or some portion of it. Consequently, our physics cannot predict (or deny) the existence of these holes. However, holes or edges that develop gradually out of 'ordinary' spacetime could be, and indeed are, predicted by sound physical principles that most physicists accept.

Figure 13 is an attempt to depict for a two-dimensional surface what a heralded edge to space – a hole with teeth – might look like. The surface is a cone-like structure that tapers gradually but relentlessly to a point known as a cusp: crudely speaking the spike is infinitely sharp, so nothing can 'turn over' the tip and climb down the other side. An object which approaches the tip starts to feel uncomfortable as the increasing curvature tries to bend it, and the diminishing room

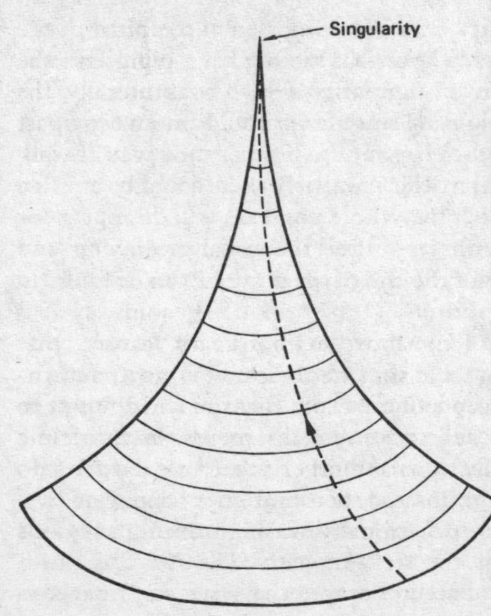

Figure 13 Hole with teeth. Space (the surface) curves progressively more until it pinches off altogether at a point, and stops. A curious observer (broken line) who explores near the tip risks disappearing for good off the end – he can never return. However, he is well warned of the impending end as he becomes violently squeezed into the diminishing spaee near the tip.

constricts it. Very near the tip the object becomes progressively squeezed, and it can only reach the tip itself by being crushed down out of existence – compressed to nothing at all – for the tip has no size. The price paid for visiting the tip is the destruction of all extension and structure; the object can never return.

These cusp-like edges to spacetime from which no traveller can return are predicted by Einstein's theory of relativity, and are known as singularities. The escalating curvature in their vicinity corresponds physically to forces of gravity, which would drag all bodies apart and smash them into an ever-decreasing volume. One way in which such an unpleasant feature might occur is from the gravitational collapse of a burnt out star. When a star's fuel is exhausted, it loses heat and cannot sustain enough internal pressure to support its own weight, so it shrinks. In rather large stars, the shrinkage becomes so rapid that it amounts to a sudden implosion and the stars contract, perhaps without limit. A spacetime singularity forms and much, maybe all, of the star could disappear into it. Even if it does not, curious observers who follow its progress can still run into the singularity. It is widely believed that if a singularity occurs, it will be located inside a black hole where one cannot see it without falling in and leaving the universe.

Another type of singularity could have existed at the birth of the universe. Many astronomers believe that the big bang represents the debris which erupted from a singularity which was literally the creation of the universe. A big bang singularity could amount to a past temporal edge to the cosmos – a beginning of time, and space as well, in addition to the origin of all matter. Similarly there could be an edge to time in the future, at which the whole universe will disappear for good – space and time with it – after the usual crunching and annihilation. Further images of the end of the universe can be found in my book *The Runaway Universe*.

Having described some of the more extraordinary features that modern physics permits space and time to possess, it is worth returning to Jiffyland and the concepts of quantum theory in an attempt to understand what the frothy substructure really means. In chapters 1 and 3 we discovered how electrons and other subatomic particles do not simply move from A to B. Instead their motion is controlled by a wave, which can spread out, occasionally washing through regions that are quite remote from the straight path. The wave is not a substance but a wave of probability: where the wave disturbance is slight (e.g. far from the straight path) the chances of finding the particle are slim. Most of the wave motion concentrates along the classical Newtonian route, which is therefore the most probable path. This bunching effect is exceedingly pronounced for macroscopic objects like billiard balls, whose wavelike spreading we never notice.

If we fire a beam of electrons (or even a single electron) from a gun, we can write down a mathematical expression for the wave, which moves according to the famous Schrödinger equation. The wave displays the important wavelike property of interference so if, for example, the beam strikes two slits in a screen, it will pass through both and the bifurcated disturbance will recombine in a structured pattern of peaks and troughs. The wave describes not one world, but an infinity of worlds, each containing a different path. These worlds are not all independent – the interference phenomenon shows that they overlap each other and 'get in each other's way'. Only a direct measurement can show which of this infinity of potential worlds is the real one. This raises delicate and profound issues of what is meant by 'real' and what constitutes a measurement, questions which will be thoroughly discussed in the coming chapters, but for now we merely note that when a physicist wishes to describe how an electron moves, or in general how the world changes, he deals with the wave and examines its motion. It is the wave which encodes all the available information about the electron's behaviour.

If we now picture all the possible worlds – say, each with a different electron trajectory – as a sort of gigantic, multi-dimensional superworld, in which all the alternatives are placed in parallel on an equal footing, then we can regard the world which is found to be 'real' on observation to be a three-dimensional projection from, or section through, this superworld. To what extent the superworld can be regarded as actually existing will be mentioned in due course. Basically we need a different world for each electron path, which usually means that we need an infinity of them, and similar infinities of worlds for every atom or subatomic particle, every photon and every graviton in existence. Clearly this superworld is a very big world indeed with infinite dimensions of infinity.

The idea that the world we observe might be a three-dimensional slice through, or projection of, an infinite-dimensional superworld may be hard to grasp. A humbler example of a projection may help. Consider an illuminated screen used to project the silhouette of a simple object, such as a knobbly potato. The image on the screen gives a two-dimensional projection of what is really a three-dimensional shape; i.e. the potato. By reorienting the potato, an infinite variety of silhouette shapes can be obtained, each representing a different projection from the larger space. Likewise, our observed world is shaped as a projection from the superworld – which projection being a matter of probability and statistics. At first sight it might seem that reducing the world to a sequence of random projections is a recipe for chaos, each successive moment presenting our senses with a completely new panorama, but the dice are heavily loaded in favour of the well-behaved, law-like Newtonian changes, so that the jerky fluctuations, which undoubtedly exist, are safely buried among the microscopic recesses of matter, only manifesting themselves on a subatomic scale.

Just as a Newtonian particle moves in such a way as to minimize its action, and a quantum wave bunches along the same path of least activity, so when it comes to gravity we find that space also conserves its activity. The quantum froth of Jiffyland fuzzes out the minimal motion somewhat, but only on the absurdly small scale discussed in the earlier part of this chapter. Thus, space itself must be described by a wave, and this spacewave will display interference properties too. Moreover, in the same way that we may construct a different world for each electron trajectory, so we may construct a different world for each shape of space. Stitching them altogether gives us an infinite-dimensional *superspace*. Contained in superspace are all the possible spaces – doughnuts, spheres, spaces with wormholes and bridges – each with a different froth arrangement; an infinity of geometric and topological arrangements and rearrangements. Each space of superspace will contain its own superworld of all possible particle arrangements. The world of our senses is apparently a single, three-dimensional element projected out of this stupendously infinite superspace.

We have now moved so far from the commonsense view of space and time that it is worth pausing to take stock. The route to superspace is a hard one to tread, each step requiring the abandonment of some cherished notion or the acceptance of an unfamiliar concept. Most people regard space and time as such fundamental features of experience that they do not question their properties in any way. Indeed, space is frequently envisaged as completely devoid of properties – an empty, featureless void. The hardest concept to accept is that space can have shape. Material bodies have shape *in* space, but space itself seems more like a container than a body.

Throughout history there have been two schools of philosophy concerning the nature of space. One school, of which Newton himself was a member, taught that space is a substance which not only has a geometry, but can also display mechanical features. Newton believed that the force of inertia was caused by the reaction of space on an accelerating body. For example, when a child whirling around on a roundabout feels a centrifugal force, the origin of this force is ascribed by Newton to the surrounding space. Similar ideas have been proposed for time, the analogy with a flowing river most closely implying an association with substance.

In contrast to these images, the alternative school proposes that space and time are not things at all, but merely relations between material bodies and events. Philosophers such as Leibniz and Ernst Mach denied that space could act on matter, and argued that all forces are due to other material bodies. Mach suggested that the centrifugal force acting on the child who rides the roundabout is caused by the relative motion between the child and distant matter in the universe. The child feels a force because the far-flung galaxies are pushing against him, resisting the motion.

According to these ideas, discussion of space and time is just a linguistic convenience enabling us to describe relations between material objects. For example, to say that there is a quarter of a million miles of space between the Earth and the moon is merely a useful way of saying that the distance from the Earth to the moon is a quarter of a million miles. If the moon were not there, and we had no other objects or light rays to manipulate, it would appear to be impossible to know how far a certain stretch of space extended. To measure distances, or angles, in space requires measuring rods, theodolites, radar signals or some other material paraphernalia. Thus space is regarded as no more of a substance than is the quality of citizenship. Both are simply descriptions of relationships that exist between things – material bodies and citizens, respectively.

Similar ideas may be applied to the concept of time. Is it necessary to regard time itself as a thing, or only a linguistic convenience for expressing the relation between events? For example, to say that one waited for a bus for a long time really only means that the interval between arriving at the bus stop and boarding the bus is uncharacteristically dilatory. The duration of time is a mode of speech describing the temporal relation between these two events.

When we approach the idea of curved spacetime, it is undoubtedly more helpful to adopt the former perspective, in which space and time are treated as substance. This may not be strictly necessary from a logical point of view, but as an aid to intuition it is helpful. Visualizing space as a block of rubber gives a vivid image of what it means for space to bend or stretch. The essential feature of Einstein's general theory of relativity is that spacetime, with this curious elastic quality, can move about, i.e. change shape, the cause of this motion being the presence of matter and energy. Once the idea of a dynamical spacetime is grasped then the quantum aspects become more meaningful.

When the concepts of quantum theory are applied to spacetime itself, the unfamiliarity is compounded because one is elaborating the already bewildering structure of a dynamical spacetime with the weird features of quantum theory. Quantum mechanics implies that we must consider not one spacetime, but an infinity of them, with different shapes and topologies. These spacetimes all fit together after the fashion of waves, each interfering with the other. The strength of the wave is a measure of how probable it is that a space of that particular shape is found to represent the actual universe when an observation is made. The spaces will evolve, such as when the universe expands, and the overwhelming number of these alternative worlds will expand in a very similar way. Some of them, however, fluctuate far from the main path, like the children in the park discussed in connection with Figure 3. The wave strength in these maverick worlds is very low, so there is only an infinitesimal chance that they will actually be observed. Down at the scale of Jiffyland, these fluctuations become far more pronounced, and random departures from smooth, unruffled space frequently occur.

Facing up to the existence of a superspace in which myriads of worlds are stitched together in a curious overlapping, wavelike fashion, the concrete world of daily life seems light years away. With concepts so abstract and disturbing as these, one is bound to wonder to what extent superspace is 'real'. Do these alternative worlds actually exist, or are they just terms in some mathematical formula that is supposed to represent reality? What is the meaning of the mysterious waves that regulate the motion of matter and spacetime alike and which define the probabilities for the existence of any particular world? What is 'existence' anyway in such a quagmire of insubstantial concepts? Where do we – the observers – fit into this scheme? These are some of the questions that we will turn to next. We shall see that the cosmic game of chance is far more subtle and bizarre than mere roulette.

6. *The nature of reality*

So far we have deliberately been rather cavalier about notions like 'the real world' and the 'existence' of matter waves, or superspace. In this chapter we shall face up squarely to the fundamental questions raised by the quantum revolution, and examine to what extent these unfamiliar concepts are supposed to apply to something truly objective, or whether they are only physicists' elaborate concoctions for computing mathematically the results of measurements of more concrete and familiar entities.

It should be stressed right at the outset that there is by no means unanimous agreement among physicists, let alone philosophers, either on the nature or existence of reality, or even its very meaningfulness, or to what extent quantum features undermine it. Nevertheless certain problems and paradoxes have been tossed around for fifty years or so, and although they are not resolved to everyone's satisfaction, they highlight the profoundly strange qualities that quantum theory has brought into our world.

Most ordinary people have an intuitive picture of reality along the following lines. The world is full of things (stars, clouds, trees, rocks . . .), included among which are conscious observers (people, dolphins, Martians? . . .) quite independently of whether they are, or have been, discovered, or whether we might plan to experiment upon them or measure them in the future. In short: there is a world 'out there'. In daily life we do not question this belief. Mount Everest and the Andromeda nebula surely existed before there were any people around to comment on the fact; electrons buzzed around the primeval universe irrespective of whether humans eventually appeared in the cosmos, and so on. Because scientists have revealed, and believe in, laws of nature, they accept that the universe 'ticks over' on its own, unaided by, and oblivious of, our own involvement in it. The obviousness of all this is all the more striking when we discover how ill-founded it is.

Clearly the world that a person actually experiences cannot be totally objective, because we experience the world by interacting with it. The act of experience requires two components: the observer and the observed. It is the mutual interaction between them that supplies our sensations of a surrounding 'reality'. It is equally obvious that our version of this 'reality' will be coloured by our model of the world as constructed by previous experience, emotional predisposition, expectation and so on. Clearly, then, in daily life we do not experience an objective reality at all but a sort of cocktail of internal and external perspectives.

The purpose of physical science has been to disengage from this personalized and semi-subjective view of the world and to build a model of reality which is *independent* of the observer. Traditional procedures to attain this goal are repeatable experiments, measurement by machine, mathematical formulation, etc. How successful is this objective model provided by science? Can it actually describe a world which exists independently of the people who perceive it?

Before discussing quantum theory, it is interesting to return to the ideas of Newtonian mechanics, with its images of a clockwork universe inhabited by observers who are mere automata, to see how far one can go in constructing a model of such a world. In chapter 3 we found that no observation can be made without some disturbance to the system being observed. To acquire information about something necessitates that some sort of influence travels from the system of interest to the brain of the observer, perhaps via a complex chain of apparatus. This influence always reacts back on the system according to Newton's law of action and reaction, thereby disturbing slightly its condition. An example was quoted about the motion of planets in the solar system, the orbits of which are minutely, but inevitably, perturbed to an infinitesimal degree by the light with which we see them. It might be thought that the disturbances due to observation deal a death blow to the idea that the universe is a machine, but this is not so. The observer's body – brain, sense organs, nervous system, etc. – can all be considered as part of the great cosmic clockwork, so considering the total system (observer plus observed) as one big machine establishes the inevitability of the outcome of all measurements. In this Newtonian picture of the universe, observers act out predetermined roles in the play – they are simply taken along for the ride. Nor is it necessary, according to this theory, for all systems and all processes to actually be observed for them to exist: who would deny that the eclipses would take place irrespective of whether anyone is around to see them? The laws of Newtonian mechanics enable one to compute the activity of unseen bodies, from atoms to galaxies, and check the predictions by only sporadic observation. The fact that the system seems to run according to these mathematical predictions reinforces the belief that it really is 'out there', operating on its own, without the need for our continual inspection to make it tick.

A central feature of this Newtonian view of a real world is the existence of identifiable 'things' that can consistently be ascribed intrinsic attributes. In daily life we have no difficulty in accepting for example, a football as a football – a definite entity with fixed properties (round, leathery, hollow . . .). It is not a house or a cloud or a star. The world is perceived as a collection of distinct objects in interaction with one another. The idea is, however, only approximate. Objects are distinct so long as their mutual interaction is in some vague sense small. When a drop of liquid falls into the ocean it interacts strongly with the larger body of water and becomes absorbed into it, losing its identity completely. To take another example, a foetus only gradually acquires a separate identity from the mother as it grows in the womb. Generally speaking, when objects are separated by a large distance, we think of them as being distinct: the planets of the solar system, the atoms in London and New York, etc. This is because all known forces of interaction diminish rapidly with distance, so that well-separated entities behave almost independently. They are never, of course, completely independent – there is always a residual coupling between all things – but the concept of distinct, separate objects is a very useful one in practice.

There is a philosophical difficulty about attributing identities to things, such as whether the football is the same football at all times. When it is kicked it loses some leather, gains some mud and boot polish, expels some air, acquires momentum and spin, and so on. Why do we think of a kicked ball as *the* ball? Similarly, it is common practice to attribute fixed identities to people, yet every day some of their body cells are replaced, and their personalities, emotions and memories are altered by the new experiences of the preceding twenty-four hours. They are not exactly the same people as we knew yesterday. At a still more fundamental level, the observed football cannot be precisely the same as the unobserved one, because of the disturbance due to the very act of observation.

The solution to these difficulties seems to be that the universe as a whole is really an inseparable thing, but to very good approximation we may divide it up into lots of quasi-autonomous little things whose separate identity, whilst open to philosophical dispute, is rarely in question in daily life. Whether or not the cosmos is considered as a single machine, or a collection of loosely coupled machines, its reality seems to be on a fairly firm footing as far as Newtonian physics is concerned. Although we are embedded in this reality, we think of it as independent of ourselves and existing before and after our own existence.

It should be mentioned that this view of reality has been criticized by a school of philosophers known as Logical Positivists, who believe, loosely speaking, that statements about the world that cannot be verified by human beings are meaningless. For example, to assert that eclipses took place before there was anyone around to see them is considered a meaningless statement. How could its truth ever be verified? Reality, for the extreme positivist, is merely what is actually perceived: there is no external world existing independently of the observer. Although it may be conceded that the reality of unobserved events cannot be established by any operational means, neither, for that matter, can their unreality. Both notions must be regarded as devoid of meaning. The positivist view of the world, at least in its extreme form, does not accord with the commonsense view, nor do many scientists subscribe to its basic tenets. Moreover, it faces its own philosophical objections (for example, how can we anyway verify the statement that unverifiable statements are meaningless?). In what follows we shall assume that some notion of an external world, independent of ourselves, has meaning, and that things exist even though we may not happen to know about them.

Turning now to quantum theory, we can already glimpse some of the problems that arise concerning the nature of reality. Whereas an observed football is only infinitesimally different from an unobserved football, when it comes to subatomic particles the act of observation has drastic effects. As pointed out in chapter 3, any sort of measurement carried out on an electron, for example, is likely to result in a large and uncontrollable recoil. However, the fact of inevitable disturbance as such does not undermine the notion of reality; but that there is no way of knowing, even in principle, the details of that disturbance. It is not possible, for instance, to attribute both the properties of a precise location and a precise motion simultaneously to a single electron. There is also a profound difficulty connected with the ability to attribute an independent existence to individual members of a collection of subatomic particles. Because all electrons are intrinsically identical, when they approach closely it may not be possible to tell which is which, for their locations may be more uncertain than their distance from each other. Neither, as we discussed in chapter 3, is it always possible to say through which slit in a screen an electron or photon 'really' passes. In spite of this, it might be supposed that one could envisage a microworld in which electrons and other particles 'really' occupy certain positions and move in well-defined ways, even though we are unable to ascertain what they are in practice. At first sight it seems that the all-important uncertainty is actually introduced by the measurement act itself, as if in some way the apparatus which is used to probe the microsystem inevitably jiggles it a bit. This was indeed the way in which the idea was introduced on page 60, but we shall see below that such a simple idea cannot be so. It is in any case clear that the jiggling effect must continue to operate even without our direct interference, otherwise all the atoms that are not directly observed would not obey the quantum laws and would collapse.

It is still possible to conjure up a picture in which all the subatomic particles really do occupy a certain position and have a definite speed, even though they are shaking about. After all, we know that the molecules of a gas, for example, are in rapid, agitated motion, an activity which is the cause of gas pressure. It is impossible for us to follow the complicated manoeuverings of billions of tiny molecules, so for practical purposes there is a deep uncertainty about how individual gas molecules will behave. This indeterminacy in molecular motions is merely due to our ignorance of their precise condition, and is similar to the heads/tails uncertainty discussed in chapter 1. In such circumstances, scientists have little choice but to use statistical methods, for although the career of any individual molecule may be quite uncertain, the average properties of a large collection can still be examined, just as the perambulatory habits of park visitors display collective order in spite of individual uncertainty (see page 30). Thus, one can accurately compute the probability of heads or tails, or the likelihood that two dissimilar gases will have diffused together after one minute, etc. Such a description of systems which are composed of random and chaotic elements, in terms of probability, seems to be very close to the quantum description of individual subatomic particles moving in probabilistic fashion. It is therefore natural to wonder whether the unpredictable behaviour of, say, an electron, owes its origin to phenomena similar to those that make the flipped coin or box of gas uncertain in their gross behaviour. Could it be that the electron and its subatomic colleagues are not really the smallest level of physical structure at all, but are themselves subject to ultra-microscopic influences that jiggle them around? If this is the case, quantum uncertainty would simply be attributed to our ignorance of the precise details of this substratum of chaotic forces.

A number of physicists have tried to construct a theory of quantum phenomena based on this idea, in which the apparently capricious and random fluctuations of microsystems do not represent an intrinsic indeterminacy in nature, but are simply manifestations of a hidden level of structure in which complicated, but fully deterministic micro-forces jiggle the electrons and other particles about. The indeterminacy of quantum systems then has the same sort of origin as that of the weather, which can only be forecast on a probability basis (e.g. there is a fifty per cent chance of rain tomorrow) and discussed in general terms with the aid of statistics.

There are two reasons why this rather obvious explanation of quantum indeterminacy has not received wide acclaim. The first is that it necessarily introduces a lot of complication into the theory because in addition to the electrons and other subatomic matter, we need to understand the details of these mysterious forces that jiggle the particles around. What is their origin, how do they operate, what laws do they, in turn, obey? The second reason is far more fundamental and goes to the very heart of the quantum revolution and any attempt to ascribe objective reality to the world of subatomic matter.

Much of this chapter will be devoted to analyzing the mind-boggling conclusions that seem to be inescapable when the nature of reality is scrutinized in the light of certain subatomic experiments. The most celebrated of these experiments was conceived in principle by Albert Einstein in collaboration with Nathan Rosen and Boris Podolsky, as early as 1935, but only in recent years has laboratory technology advanced to the point where their ideas can be tested. The experiments have confirmed that, at least in simple form, the possibility of quantum uncertainty arising simply from a substratum of jiggles, is not viable.

The underlying principle of the Einstein-Rosen-Podolsky 'paradox', as it became known, can be grasped by imagining a projectile being fired, say from a gun. Experience shows that the gun recoils so that the forward momentum of the bullet is exactly balanced by an equal and opposite momentum of the gun. If gun and bullet were equally massive, each would try to fly off in opposite directions at the same speed. If the bullet is now fired in such a way as to set it spinning, the same principle requires the gun to spin in the opposite direction. Both the forward and the twisting motions of the bullet react on the gun at the moment of departure to impart an oppositely directed kick.

Some subatomic particles also emit spinning projectiles and suffer recoils, and experiments show that these familiar rules of mechanics apply to their motions also. The particles may even disintegrate into two identical progeny, which fly off in opposite directions with counter-rotating spins. For example, the electrically neutral pi-meson, which has no spin, explodes in a mere ten-millionth-billionth of a second into two oppositely moving photons, one of which spins clockwise along its path, the other anticlockwise. The rules of quantum theory require it to be equally probable that the photons spin either way because, by symmetry, there is no reason why one particular rotation direction should be favoured over the other. Thus, if they appear moving north-south, then the north-moving one is equally likely to be spinning clockwise as anticlockwise. However, if the northerly one spins clockwise then the southerly one must, to comply with the above mentioned laws of mechanics, spin anticlockwise, and vice versa (see Figure 14). Because of this inescapable correlation between the spin directions of the two progeny, a measurement of

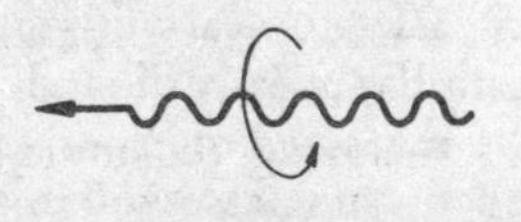

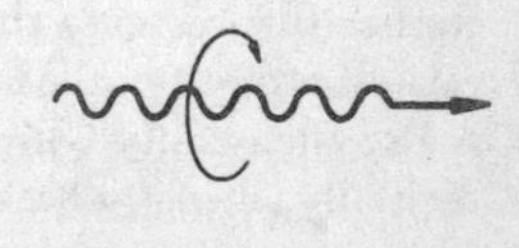

Figure 14 Spin correlation
When the neutral pi-meson decays into two photons, the spin of one must be opposite to that of the other, so if the spin of, say, the right hand photon is measured, that of the left hand photon is deduced immediately. However, paradoxes arise when it is realized that the spin direction is intrinsically indeterminate until the measurement actually occurs.

which way one of them is spinning immediately gives the information about which way the other is spinning.

The essential feature of this example is that after the disintegration of the parent body, the two product particles may move a great distance apart. Indeed if the explosion occurs in outer space, the particles might continue flying apart until separated by millions of light years. If the spin measurement is now made, a local observation of the spin direction of one particle gives immediate information about the other particle, which is far outside the galaxy. Now according to the theory of relativity, information cannot travel faster than light, so the idea of the instantaneous acquisition of knowledge about a particle in a very distant place might appear to violate a fundamental principle. In the case of the gun and bullet the common-sense picture is that, long before the observation of the spin direction was made, the bullet was 'really' spinning, say, clockwise and the gun anticlockwise, and the only effect of the measurement is to make this knowledge available to an observer. This does not really amount to sending a signal faster than light as no actual physical influence travels between the two bodies. So provided we assume the existence of a real world, independent of our knowledge or our intention to make an observation, containing real objects (guns, bullets) with fixed and meaningful attributes (spinning, separating), there is no conflict with the principles of relativity and the inability to send signals faster than light.

It is natural to extend this picture to the subatomic domain also, and to suppose that the two particles are 'really' spinning in such-and-such a way, irrespective of whether we intend to find out by performing an experiment. Thus Einstein hoped to establish the independent reality of the physical world but any straightforward attempt to claim that such entities are 'really' behaving in a certain way before we observe them has been demolished by recent experiments.

Let us choose as the two separating particles photons of light. Rather than discuss their spins, as above, it is easier to deal with a related property called polarization, for this is familiar in daily life and is also the quality which physicists have actually measured to experimentally check what will be described. Modern sunglasses often include polarizing glass, and an understanding of their operation is basically all that is needed to comprehend why the world is not as real as it may seem. Light is an electromagnetic vibration and we can ask in which direction the electromagnetic field vibrates. A mathematical study, or some simple experiments, show that if the wave is travelling, say, vertically, then the vibrations are always horizontal; the wave motion is transverse to the direction of travel. By symmetry, a random, vertical light beam would not show any preference for any particular horizontal direction to vibrate along; it could be north-south, east-west or any other choice in between. The point about polarized glass is that it is only transparent to light which vibrates in one particular direction. If we examine the light that emerges from such a polarizer, we find it is all vibrating along this special direction; so it acts as a filter which sorts out light of only one chosen vibration direction. This refined light is called 'polarized'. Naturally we can freely choose the polarization direction by rotating the polarizer.

Suppose we now bring up a second polarizer behind the first. If their two special directions are oriented parallel then all the light that passes through the first also goes through the second, because the latter accepts light with the same direction of polarization as itself. On the other hand, when the second polarizer is held perpendicular to the first no light gets through (see Figure 15). Finally, if the second polarizer is

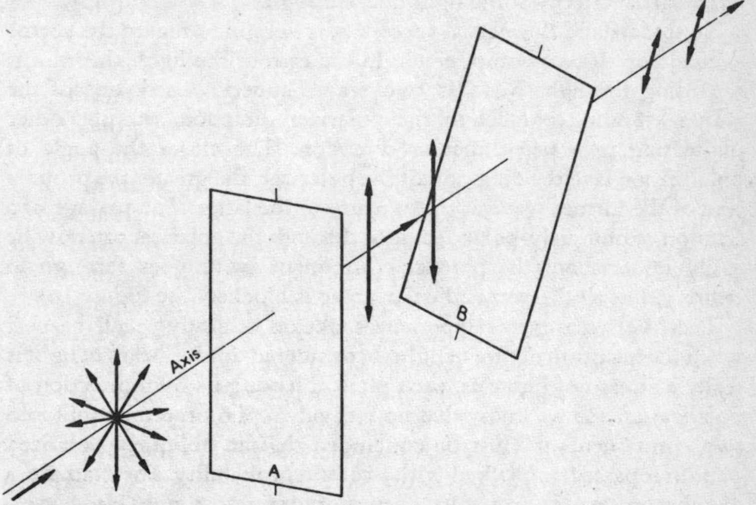

Figure 15 Polarizers
Light waves vibrate perpendicular to their line of motion. Ordinary light is a superposition of vibrations in every direction, but on emerging from the polarizer A only one direction of vibration remains. This light is called polarized. When polarized light strikes a second polarizer B obliquely, only a fraction is passed. The transparency of B depends on the orientation: if it is parallel to A, all the polarized light is passed; if it is perpendicular, none gets through.

held at some oblique angle in between these two extremes then some, but not all, the light gets through the second polarizer. This, incidentally, is why polarizers are used for sunglasses, because a lot of glare reflected from glass or water, and also some sky brightness, is partially polarized by the reflection process, so unless the sunglasses are oriented in the direction of this polarized light, they will block out a good fraction of it.

The reason that a polarizer will still accept at least a fraction of the light that is vibrating obliquely to it can be understood by the analogy with an attempt at pushing a car (see chapter 3). The light vibration is also a vector, and if it lines up with the polarizer angle then the light gets through, but if it is perpendicular, nothing happens – the light is blocked. The important feature is that one can still push a car moderately effectively by exerting an oblique force, say whilst leaning against the driver's door so as to keep control of the steering wheel. The closer the angle of push is to the line of motion, the more efficiently the car responds. Similarly, obliquely polarized light also has a partial effect – some light gets through.

To understand this partial success, it is helpful to regard the vector as made up of two components. In the case of the light, this means regarding the light wave as two waves superimposed, one of the waves vibrating parallel to the polarizer direction, and the other undulating in a perpendicular direction. The closer the angle of polarization is to the direction of the polarizer, the greater the proportion of the former wave at the expense of the latter. The passage of a fraction of obliquely-polarized light through the polarizer can now be easily understood: the parallel component wave goes through in entirety, but all the perpendicular wave is blocked (see Figure 16).

These very reasonable experiences take on a slightly peculiar aspect when the quantum nature of light is considered, for the beam of light is really a stream of photons, each photon having its own direction of polarization. As we know that no individual photon can be split into two components it must be concluded that an obliquely polarized photon is passed or blocked with a certain probability. For example, a 45° photon stands a fifty-fifty chance of getting through. However – and this point is crucial – once it has got through the photon must emerge with its polarization parallel to that of the polarizer because, as we have just seen, the light that has passed through a polarizer emerges completely polarized in the same direction.

The conclusion is that when the photon interacts with the polarizer,

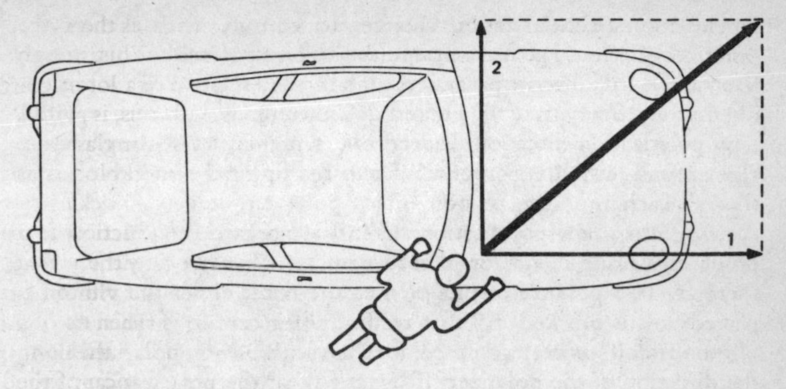

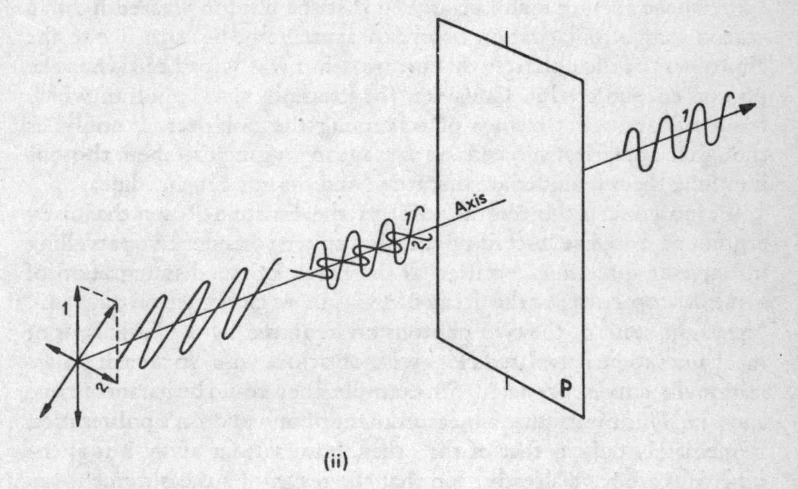

Figure 16 Resolving vectors
(i) The oblique force (heavy arrow) may be considered as two weaker forces superimposed: a component (1) directed along the road, which pushes the car, and a perpendicular component (2) which is blocked. The relative strengths of 1 and 2 depend on the angle of push. (ii) Similarly, a polarized light wave may be considered as two weaker waves superimposed, one (1) which vibrates parallel to the polarizer (P) and gets through; the other (2) is perpendicular to it and gets blocked.

its direction of polarization changes to comply with that of the polarizer. We could go on to pass it (with a certain likelihood) through a second, third, or more polarizers each twisted relative to the preceding one, and every time the photon gets through it will emerge with a new polarization direction. Indeed one can even twist the direction right round until it is perpendicular to the original direction. It is as though each time the photon hits a polarizer, it gets knocked or thrown into a new polarization state. If we regard the polarizer as a crude measuring device or photon detector, then we could say that there are two possible results of measurement: either the photon is passed or it is blocked. All that we know for certain is the state of a photon after it has been accepted, for then we *know* it is polarized along the direction of the polarizer. If we ask what the polarization of the photon is before the measurement is made, i.e. before it emerges from the polarizer, then there is a difficulty, for the polarizer has apparently disturbed the photon's state and imposed its own direction upon it. Nevertheless it might still be argued that the photon was 'really' in a certain state of polarization before measurement, but that due to the clumsiness of the polarizer that information was wiped out when the photon encountered it. Consider, for example, a 45° photon, which has a fifty per cent chance of traversing the polarizer. It seems as though the polarizer succeeds on average in jogging half these photons into line; the remainder are discarded and do not get through.

We now reach the central point of the Einstein-Rosen-Podolsky argument. Suppose instead of one photon we consider two travelling in opposite directions, emitted as the result of the disintegration of some other particle, or the decay of an atom, as explained on page 113. Just as the spins of the two photons are required by the basic laws of mechanics to be correlated clockwise-anticlockwise, so are the polarization directions correlated: for example, they could be parallel to one another. This means that a measurement of one photon's polarization immediately tells us that of the other, however far away it is at the time. But we have already seen that the result of a measurement can only be 'yes' or 'no' according to whether the photon is allowed to pass through a polarizer or not. We can only say what the photon state is *after* the measurement has taken place, i.e. when it emerges from the polarizer, and this is true at whatever angle we choose to place the polarizer. We can only detect photons in one of two states – parallel and perpendicular to the polarizer (corresponding to 'yes' and 'no'). However, the choice of *which* two states is entirely up to us; the polarizer can be oriented at whim. The truly mind-boggling implications of this freedom are apparent if we use two parallely orientated polarizers, one of which is stationed in the path of each of the two correlated photons. Because the polarizations are forced to be parallel, whatever we measure for the photon polarization of one we are obliged to find the same for the other, but as there are only really two polarization states that are measurable (i.e. parallel and perpendicular) the 'yes'-'no' decision of one polarizer must be identical to that of the other. That is, every time a photon is passed by one polarizer, the other one *must* allow its photon through also, and whenever one photon is blocked, so is the other (see Figure 17). Extraordinary though these ideas may seem, they have been carefully checked by experiment in the laboratory, and the details described here were verified.

Figure 17 Einstein-Rosen-Podolsky paradox
The atom sends two photons simultaneously towards two parallel polarizers. If A passes its photon, so does B. How does B know what A will do? A and B could be light years apart, and either A or B could pass its photon first. Bohr concluded that the photons are not truly real until they encounter the polarizers.

The profound peculiarity of this result is evident when we realize that the photons may have moved millions of miles apart by the time they hit their respective polarizers, yet they still cooperate in their behaviour. The mystery is, how does the second polarizer *know* that the first one has let a photon pass, so that it too may do the same? The experiments could be carried out simultaneously, in which case we can be sure, on the basis of the theory of relativity, that no message can travel faster than the photons themselves between the polarizers to say 'let this one pass'. In fact, by stationing the polarizers at different distances from the decaying atom we could arrange for either experiment to be performed before the other, thereby ruling out any question of one polarizer signalling the other, or causing it to accept or reject a photon. Indeed the theory of relativity enables observers in different states of motion to disagree anyway about the time order of widely separated events, so if it were alleged that polarizer A caused B to accept or reject on the basis of its own decision, someone moving differently would see B accept or reject *before* A even knew what it would do with its photon!

These observations make it clear that the indeterminacy of the microworld cannot be introduced by the measuring apparatus itself, or simple random jiggles as the photon goes on its way, for there would be no reason why the two different polarizers should cooperate in this remarkable way to block or pass their respective photons in unison. If each photon gets its polarization direction jogged around at random, there is no reason to expect them to arrive at their respective polarizers jogged into exactly the same direction. One would expect, on average, half the photons to be accepted by one polarizer when the other rejects its photon, but this is in clear contradiction to the above predictions of quantum theory and to the experiments that have verified them. The conclusion must be that subatomic uncertainty is not just a result of our ignorance about microforces but is inherent in nature – an absolute indeterminacy of the universe.

The Einstein-Rosen-Podolsky experiment has staggering implications for the nature of reality if taken at face value. A last vestige of the commonsense view of reality can only be retained if it is claimed that when both polarizers mysteriously collaborate to simultaneously accept their photons, then these photons were all along 'really' polarized exactly parallel to the polarizers, thereby ensuring their eventual passage through their respective polarizers, and that those that are blocked were 'really' always vibrating perpendicular to the polarizers. But the absurdity of this last desperate attempt to hold on to a 'real' world is not only the fact that the original atom must contrive to know at which angle the polarizers are set, but that we can change our minds about this angle after the photons have been emitted. It is scarcely conceivable that the behaviour of an atom should be influenced by our decision to experiment on a photon it emits at some stage in the future. As all other atoms happily emit photons with all manner of polarizations in a perfectly random way, our experimental whims can hardly be expected to affect a particular one, especially as we could choose to detect photons from atoms that are billions of light years away, at the other end of the universe. As if these objections were not enough, it can be shown mathematically that if the photons really were in *either* one state (parallel to the polarizers) *or* the other (perpendicular) then the 'yes'-'no' cooperation would fail. Though more complicated jiggle theories exist in which the polarizers' cooperation is explained, these have been ruled out by experiments with non-parallel polarizers.

The wave nature of quantum processes enters here in a vital way. To eliminate absurdities about atoms anticipating our experiments, let us suppose that we produce a beam of photons with a definite polarization direction by already passing them through one polarizer. When they approach another polarizer that is slanted relative to the first, they will either be accepted or rejected with a definite probability that depends in a simple arithmetical way on the angle of the slant. If this is 45°, half the photons are passed on average. Viewed on this slant, the polarized beam can be envisaged as made up of two waves of equal strength, one parallel and one perpendicular to the second polarizer. These two waves must be present *together* in order to constitute the original, unslanted polarized wave. The interference effects between the two waves play an essential role. It is not possible to say that either the parallel or the perpendicular wave alone exists, for this contradicts the fact that we know already that the wave is polarized neither parallel nor perpendicular to the second polarizer, but at 45° to it. When only one photon is involved the implications are bizarre. It is not possible to say that this photon has a polarization either parallel or perpendicular to the polarizer but, because the wave interference still exists even for one particle, *both* possibilities must co-exist and overlap each other. Moreover, the angle of the polarizer, and hence the relative admixture of the two alternatives, is entirely in the control of the experimenter! It must be emphasized that quantum indeterminacy does not merely mean that we cannot know which polarization direction the photon really possesses, it means that the notion of a photon with a definite polarization direction does not exist. There is an inherent uncertainty in the *identity* of the photon itself, not just in our knowledge of it. Similarly, when it is said that we are uncertain about the location of an electron it is not simply that the electron *is* at one place or another, which we cannot ascertain. The uncertainty concerns the very identity of the 'electron-at-a-place'.

In the spirit of the superspace idea, we can regard the two photon waves as representing two worlds, one in which the second polarizer accepts the photon, one in which it is rejected. Moreover, these two worlds could be very different in character, for the accepted photon could go on to trigger a detonator which explodes a hydrogen bomb. However – and this is the culmination of the long analysis given in this chapter – these two worlds are not independent realities. They are not 'either-or' alternative worlds; they *overlap* each other. That is, the crucial interference effects caused by the overlapping of the two waves shows that, before the second polarizer decides on the fate of the photon, *both* these worlds are mixed together. Only when the polarizer finally decides do the two worlds become separate choices for 'reality'. The effect of the measurement by the second polarizer is to chop the overlapping worlds apart into two disconnected alternative realities.

We have now arrived at some idea of the nature of reality according to the usual interpretation of quantum theory, but it is a pale shadow of the commonsense image. The indeterminacy of the microworld is not just a consequence of our ignorance (as with the weather) but is absolute. We are not merely presented with a choice of alternatives, such as the heads/tails unpredictability of daily life, but a genuine hybrid of the two. Until we make a definite observation of the world it is meaningless to ascribe to it a definite reality (or even various alternatives), for it is a superposition of different worlds. In the words of Niels Bohr, one of the founders of quantum theory, there are 'fundamental limitations, met with in atomic physics, of the objective existence of phenomena independent of their means of observation'. Only when the observation is made does this schizophrenic state collapse onto something that is in any sense real.

In the previous chapter it was explained how the world we observe is a slice through, or projection of, an infinite dimensional superspace – a vast collection of alternative worlds. We now see that the world we observe is not just a random selection from superspace, but depends in a crucial way on all the other worlds that we don't see. Just as the 'yes'–'no' correlation between the two separated polarizers depends crucially on the interference between the 'yes' world and the 'no' world, so in every other interaction, in every far-flung atom, every microsecond, all the worlds-that-never-were leave a vestige of their putative reality in our own world by controlling the probabilities of all these subatomic processes. Without the other worlds of superspace, the quantum would fail and the universe would disintegrate; these countless alternative contenders for reality help steer our own destiny.

According to these ideas, reality only makes sense within the context of a prescribed measurement or observation. We cannot generally say that an electron, or a photon, or an atom was really behaving in such and such a way before we measured it. The only reality is the total system of subatomic particles plus apparatus and experimenter, for if the experimenter chooses, for example, to rotate his polarizer, then he changes the choices for the alternative worlds.

Every time someone wearing polaroid sunglasses nods his head, he rearranges the selection of worlds in superspace. He can choose whether to create a world of north-south photons, east-west photons, or whatever else takes his fancy.

It follows that the observer is involved in reality in a very fundamental way: by choosing the experiment he chooses which alternatives are on offer. When he changes his mind, he changes the selection of possible worlds. Of course, the experimenter cannot pick precisely which world he wants, for they are still subject to the rules of probability, but he can influence the available choice. In short, we cannot load the dice, but we can decide the game we want to play.

We can now accept that this involvement of the observer in his own reality is far more profound than the classical Newtonian picture of the world in which the observer is embedded in reality, but only as an automaton, whose acts are completely determined by the laws of mechanics. In the quantum version there is an inherent indeterminism, and the particular reality only appears within the context of a particular type of measurement or observation. Only when an experimental set-up has been specified (e.g. which angle the polarizer is chosen to have) can the choices of reality be specified. Some scientists have suggested that by discrediting the Newtonian idea of a mechanistic universe inhabited by observers who are mere automata, quantum theory restores the possibility of free will. If the observer in a sense chooses his own reality, does that not amount to a freedom of choice and an ability to restructure the world according to our own whims? Though the answer may be positive, we must remember that in quantum theory the observer (or experimenter) cannot in general determine the outcome of any particular experiment. As already emphasized, the only choice we have is over the various alternative outcomes, not over which alternative is realized. Thus, we can decide to create a world in which some photons are either polarized north-south or east-west, or else another world in which they are polarized either northeast-southwest or northwest-southeast, etc. However, we cannot choose which of the two possibilities will occur in each case. We cannot force a randomly-polarized photon to be north-south rather than east-west, because we cannot force it to pass through a north-south oriented polarizer. Similarly, we may choose to measure either the position, or the momentum of a particle, but not both. After the measurement the particle will then have a well-defined value of one or the other – depending on our choice of experiment.

We seem to have a situation in which the universe is in a sort of suspended state of schizophrenia until someone undertakes an observation, when it 'collapses' suddenly into reality. Moreover, as the earlier lengthy discussion about two correlated photons travelling in opposite directions has underlined, the collapse into reality occurs not just locally (e.g. in the laboratory) but also suddenly and instantaneously in distant regions of the universe. From the theory of relativity it is known that different observers generally disagree about what is instantaneous, so that the onset of reality appears to be entirely a personal affair. Consequently it is not possible to use this collapse as a signalling device between distant observers.

According to relativity, any signal sent faster than light would threaten causality, for it would then not only be possible to send a signal backwards in time from the point of view of another observer, but even to signal one's own past. This possibility raises horrendous paradoxes concerning 'autocidal' machines that are programmed to self-destruct at two o'clock if they receive a signal at one o'clock transmitted by them at three o'clock. If they are destroyed at two they cannot transmit at three, so no signal is received and no destruction occurs. But if no destruction occurs the signal *is* sent so destruction *does* occur. This obvious contradiction seems to rule out reversed-time signalling, and hence faster than light messages.

In the quantum case, we have seen how the passage of a photon through a polarizer in one place can ensure the passage of another photon through another polarizer somewhere else, perhaps millions of miles away, at the same moment (relative to that particular experimenter) or, indeed, even *before* that moment. In spite of this amazing property, the experimenter has no control over each individual photon, because of quantum uncertainty, so he cannot arrange with a distant colleague that, for example, the passage of three consecutive photons through the polarizer means that Everton have won the FA Cup. Therefore, the theory of relativity remains unscathed and the possibility of rapid superluminal signalling across the universe, with its attendant threat to causality, is illusory.

Although distant systems, such as our two photons and polarizers, cannot be linked by a conventional type of communication channel, neither can they be considered as separate entities. Even though the two polarizers may be in different galaxies, they inevitably constitute a single experimental arrangement, and a single version of reality. In the commonsense view of the world we regard two things as having separate identities when they are so far apart that their mutual influence is negligible. Two people, or two planets, for example, are regarded as distinct things, each with its own attributes. In contrast, quantum theory suggests that, at least before an observation is made, the system of interest cannot be regarded as a collection of things but as an indivisible, unified whole. Thus the two distant polarizers and their respective photons are not actually two isolated systems with independent properties, but linked enigmatically through the quantum processes. Only after the observation is made can the distant photon be regarded as acquiring a separate identity and an independent existence. Moreover, we have seen how it is meaningless to assign properties to a subatomic system in the absence of a precise experimental arrangement. We cannot say that a photon 'really' has such-and-such a polarization before measurement. It is therefore incorrect to regard the polarization of a photon as a property of the photon itself; rather, it is an attribute which must be assigned to both the photon and the macroscopic experimental arrangement. It follows that the microworld only has properties by *sharing* them with the macroworld of our experience.

The real challenge to our commonsense view of reality comes when the atomic nature of all matter is taken into account. We might feel that results of obscure experiments involving polarized photons bear little relevance to our daily lives, yet all the familiar things around us – all material bodies – are composed of atoms, subject to the laws of quantum theory. In a thimbleful of ordinary matter there are some thousands of billions of billions of atoms, each colliding with one another millions of times a second. According to the ideas outlined here, when two microscopic particles interact and separate they can no longer be considered as independently real things, but are correlated, though usually in ways much more complicated than the two photons involved in our discussion. It follows that all across the universe, quantum systems are coupled together in this strange fashion into a gigantic, indivisible assembly. The original belief of the ancient Greeks that all matter is made up of individual, independently existing atoms seems to be a gross oversimplification, for the atoms themselves are not individually real. Only in the context of our macroscopic observations does their reality have meaning. But our observations are exceedingly limited, both to the gross features of matter – for we rarely observe the individual atoms except in special experiments – and to our immediate corner of the universe. So we arrive at a picture in which the vast majority of the universe cannot be considered as real at all, in the traditional sense of the word. Indeed, John Wheeler has gone as far as to claim that the observer literally creates the universe by his observations:

> Is the very mechanism for the universe to come into being meaningless or unworkable or both unless the universe is guaranteed to produce life, consciousness and observership somewhere and for some little time in its history-to-be? The quantum principle shows that there is a sense in which what the observer will do in the future defines what happens in the past – even in a past so remote that life did not then exist, and shows even more, that 'observership' is a prerequisite for any useful version of 'reality'.

Needless to say, these radical ideas concerning reality which are embodied in the quantum theory have caused decades of controversy and debate. While there is little doubt that at the operational level the theory is a brilliant success – physicists are in no doubt about how to actually calculate the properties of atoms, molecules and subatomic matter using the theory – nevertheless the epistemological and metaphysical aspects of quantum physics continue to cause anxiety. The interpretation described in this chapter is due chiefly to Niels Bohr, who was one of the originators of quantum theory. It is usually known as the Copenhagen interpretation after Bohr's group in Denmark, and is probably the one to which most physicists subscribe. However, some have found the ideas it contains paradoxical, meaningless or incomplete. Albert Einstein, in particular, thought the theory incomplete because he could not see how a distant photon and polarizer could be induced to respond according to the behaviour of a nearby photon and polarizer. How can the distant one 'know' it is to accept or reject a photon without some elaborate signalling mechanism which would necessarily violate the principles of Einstein's own theory of relativity, by being superluminal?

In reply to Einstein's challenge, Bohr maintained that microscopic systems have no intrinsic properties whatever, so that it is unnecessary to consider the condition of one photon being signalled to another, for each photon in isolation does not have a meaningful condition anyway. Only the total experiment has meaning. Bohr proposed that the only true reality is that which can be communicated in plain language between people, such as the description of a click on a Geiger counter or the passage of a photon through a polarizer. Any discussion of what a photon, or an atom, etc. is 'really' doing must be approached only through the framework of an actual, concrete experimental set-up. Referring to these experimental conditions, which determine the sort of properties that can be measured, Bohr claimed that they 'constitute an inherent element of . . . physical reality'. He thus circumvented Einstein's objections.

In spite of the popular appeal of the Copenhagen interpretation and Bohr's skilful arguments, some physicists continue to find the ideas involved paradoxical because they base reality on classical concepts of experimental apparata which are themselves discredited by quantum theory. Classical, Newtonian physics – the physics of plain language and everyday, commonsense objects that Bohr wishes to use – is known to be wrong. To make use of plain language to define microscopic reality thus seems to be inconsistent. In the next chapter we shall see how some alternative interpretations of quantum theory have been proposed which are still more fantastic in their implications.

7. *Mind, matter and multiple-worlds*

We have seen how quantum physics has undermined the common-sense notion of objective reality and has placed the observer and his experiments in a central position in the definition of any proper concept of a real world 'out there'. However, there is still some vagueness about what precisely constitutes an 'observer' and what sort of physical process is involved in his 'observation'. The Copenhagen interpretation makes great use of the 'experimental apparatus'. What exactly is this?

A realistic laboratory is equipped with many devices for probing the structure of atoms and their constituents. Some of these are familiar: x-ray machines, Geiger counters, bubble chambers, high energy particle accelerators and photographic plates. However, all these bits of apparatus, not to mention laboratory technicians, are made up of atoms, and even Bohr conceded that they too must be subject to the minute uncertainties that are the characteristic of quantum physics. There is no clear dividing line between what is a microscopic system and what is a macroscopic measuring device. Quantum processes can be observed in molecules containing many atoms, and can even become prominent in visible quantities of fluids and metals. The phenomenon of superconductivity, for example, in which the electrons in a metal combine in pairs and then cooperate on a macroscopic scale to produce completely resistanceless electric current flow, is an example of quantum effects on an engineering level. Clearly it is not possible to point to something and say 'that is microscopic, and subject to quantum theory' and 'that is macroscopic, and subject to classical, Newtonian physics'.

If all systems are ultimately quantum in nature, a paradox seems to surround the act of measurement. To focus ideas, let us take a simple type of observation involving a radioactive atomic nucleus. Such a nucleus will emit one or more subatomic particles which can sometimes be detected in a Geiger counter: if the counter clicks the nucleus has decayed, if it does not, the nucleus is intact. Rather than a click, some counters are equipped with pointers that flick over a scale: if the pointer rests at position A, the nucleus is intact, if it jumps to, say, position B, a particle has been detected, and we can infer that the nucleus has decayed. Therefore, the condition of the pointer is correlated with the condition of the nucleus in a simple way. By observing the pointer we can effectively observe the nucleus.

All measurements involve the twin elements described here, which are an indispensible part of the observation process, i.e. correlation between the microscopic condition of the system of interest with some macroscopically distinguishable states of the apparatus, and amplification of the minute quantum effects to produce a large-scale change of some sort, such as the deflection of a pointer. According to quantum physics the state of the microscopic system must be described by a superposition of waves, each wave representing a definite value of some quality, such as position, momentum, spin or polarization of a particle. It is vital to remember that the superposition does not represent a set of alternatives – an either/or choice – but a genuinely overlapping combination of possible realities. The actual reality is determined only when the measurement of these qualities is effected. Here, however, is the rub. If the measuring device is also made of atoms, it too must be described by a wave which is made up of a superposition of all its alternative states. For example, our Geiger counter is in a superposition of states A and B (undeflected and deflected pointer) which, it must be repeated, does not mean *either* it is deflected *or* it is undeflected, but in some strange, schizophrenic way, *both*. Each represents an alternative reality generated by the decay of the nucleus, but these realities not only co-exist, they overlap and interfere with each other by the wave interference phenomenon.

The reason that we do not notice the overlapping of 'other realities' with our own is because in a chunk of laboratory-sized apparatus the interference effect is almost infinitesimally small. Whereas inside atoms the alternative worlds jostle with each other in a vigorous way, on an everyday scale their mutual influences are almost non-existent. But not quite. If we really believe that quantum theory applies to macroscopic objects then we have to concede that, however small, these influences from overlapping realities are invading our world. With such deep issues of principle at stake the smallness of the effect is hardly relevant, for in principle we could detect this interference with additional extremely complicated and delicate apparatus.

Until now we have had a picture of the universe as a superposition of overlapping realities in superspace, which are chopped apart into disconnected, alternative worlds as soon as an observation is made. Now we see that the chopping mechanism is not quite completely effective, and some minute threads continue to connect our world with the other worlds of superspace. The chopping can only be complete, and reality become fully objective, when a truly non-quantum device is used for measurement, otherwise there will always be residual interference between different worlds. But do any truly non-quantum systems exist? If they do it is probable that they could be used to violate the rules of quantum theory, if they don't it seems that there can be no reality. How can we escape this dilemma?

In the 1930s the mathematician John von Neumann investigated the quantum measurement process in great detail. He argued mathematically that when a microscopic system is coupled to a macroscopic measuring device, the effect of the coupling is to cause the micro-system to behave subsequently as if the interference effects are absent. That is, the state of the microsystem seems to collapse from a superposition of overlapping states, to a genuine either-or set of alternative possibilities. Unfortunately his analysis does not amount to a demonstration of 'collapse' into reality, because another outcome of the coupling is to transfer the interference effects into the measuring apparatus itself, and in order for the apparatus to 'collapse' into reality, a further system must make another measurement of the apparatus. But the same reasoning may then be extended to that further system, requiring yet another device to measure that device, and so on, apparently ad infinitum.

Where does this chain end? Erwin Schrödinger, who invented the wave theory of quantum mechanics, called attention to a curiosity that has become known as the cat paradox. Suppose our microsystem consists of a radioactive nucleus that may or may not decay after, say, one minute, according to the laws of quantum probability. The decay is registered by a Geiger counter, which is in turn attached to a hammer, in such a way that if the nucleus decays and produces a response in the counter, it releases a trigger which causes the hammer to fall and break a cyanide capsule. The whole assembly is put into a sealed box along with a cat. After one minute there is a fifty per cent chance that the nucleus has decayed. The device is switched off automatically at this stage. Is the cat alive or dead?

The answer would seem to be that there is a fifty-fifty chance of finding the cat alive when we look in the box. However, if we follow von Neumann and agree that the overlapping waves which represent the decayed and intact nucleus are correlated with overlapping waves describing the cat, then one cat-wave corresponds to 'live-cat' the other to 'dead-cat'. But these waves are both present, and interfering (minutely) with one another. The state of the cat after one minute cannot be *either* 'alive' or 'dead' because of this overlap. On the other hand, what sense can we make of a 'live-dead' cat?

On the face of it, it seems that the cat goes into one of the curious states of schizophrenia that was much discussed in the previous chapter, and its fate is only determined when the experimenter opens the box and peers in to check on the cat's health. However, as he can choose to delay this final step as long as he pleases, the cat must continue to endure its suspended animation, until either finally dispatched from its purgatory, or resurrected to a full life by the obliging but whimsical curiosity of the experimenter.

The unsatisfactory aspect of this description is that the cat itself presumably knows whether it is alive or dead long before anyone looks in the box. It might be argued that a cat is not a proper observer, inasmuch as it does not possess the full awareness of its own existence that humans enjoy, so would be too dim-witted to know whether it was alive, dead, or alive-dead. To circumvent this objection we could replace the cat with a human volunteer, sometimes known to the physics fraternity as 'Wigner's friend', after the physicist Eugene Wigner who has discussed this aspect of the paradox (see Figure 18). With such a capable accomplice installed in the box we can, if we find him alive at the end of the experiment, ask him what he felt during the period before the box was opened. There is no doubt that he would answer 'nothing', in spite of the fact that his body was supposed to have been in a live-dead state for the duration of the experiment, after which it dramatically collapsed into a living condition once more. It is true that people sometimes complain of feeling half dead, but it is hard

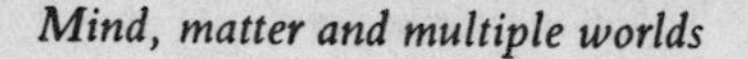

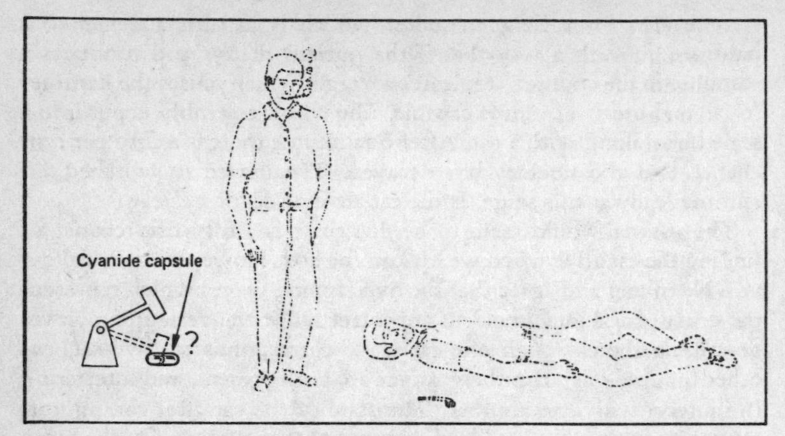

Figure 18 The paradox of Wigner's friend
The radioactively-controlled hammer can break the cyanide capsule with a certain likelihood. So is the man in the box alive or dead? Quantum theory says both – the two worlds of superspace coexist and overlap. Only when Wigner looks in the box do the worlds decouple and one of them collapses into reality. But what did his friend feel like in this schizoid condition of unreality before Wigner peeked?

to imagine that quantum interference phenomena have much to do with this particular condition.

If we insist on adhering to quantum principles at all costs then we are driven into solipsism – the conclusion that the individual (in this case the reader) is the only one who really exists, all others being unconscious robots merely forming part of the scenery. If Wigner's friend is a robot, he cannot be relied upon to report truthfully his observations, for he does not really experience any. Now this is a big step, for it pitches the observer into the centre of reality in a way that is still more crucial than we have already come to accept. To escape from solipsism, Wigner himself has proposed that quantum theory cannot be correct under all circumstances; that when the conscious awareness of the observer is involved the theory breaks down and the description of the world as a set of overlapping waves is invalidated. Solipsism has attracted its share of adherents over the centuries, but most people find it unpalatable, including Wigner. In Wigner's interpretation of quantum theory, the minds of sentient beings occupy a central role in the laws of nature and in the organization of the universe, for it is precisely when the information about an observation enters the consciousness of an observer that the superposition of waves actually collapses into reality. Thus, in a sense, the whole cosmic panorama is generated by its own inhabitants! According to Wigner's theory, before there was intelligent life, the universe did not 'really' exist. This places a grave, indeed cosmic, responsibility on living things to sustain the existence of everything else, for if all life were to cease, all the other objects – from every distant star to the smallest subatomic particle – would no longer enjoy an independent reality, but would lapse into the limbo of superposition. The bonus gained from this awesome role is that Wigner's friend-in-the-box can now bring about the collapse into reality of the box contents – himself included – so that when Wigner eventually opens the box and asks him how he felt a few moments before, he can announce 'fine', secure in the knowledge that he was one hundred per cent real already, without enlisting the help of Wigner's tardy observation of his condition to collapse his body and mind into reality.

Wigner's idea has, not unexpectedly, been widely criticized. Consciousness is normally regarded by scientists as at best ill-defined, (is a cockroach conscious? a rat? a dog? . . .) and at worst non-existent physically. Yet it has to be conceded that all our observations, and through them all of science, are based ultimately on our consciousness of the surrounding world. As usually conceived, consciousness can be acted upon by the external world, but cannot itself act on the world, thus violating the otherwise universal principle that every action induces some reaction. Wigner proposes to reinstate that principle in the case of consciousness also, so that it may react on the world by, in fact, collapsing it from a superposition into reality.

A more serious objection to Wigner's ideas is exposed if two observers become involved in observing the same system, for then each has the power to collapse it into reality. To illustrate the sort of problems that can arise, suppose that we again consider a radioactive nucleus, the decay of which will trigger a Geiger counter, but this time there is no conscious observer immediately involved. It is arranged that after one minute, when the chance of a decay is fifty per cent, the experiment is terminated and the Geiger counter pointer is locked in whatever position it held, i.e. deflected if the nucleus decayed, undeflected if it had not, so that it can be read at any time thereafter. Rather than an experimenter looking directly at the pointer, the Geiger counter is photographed. When eventually the photograph is developed, the experimenter looks at it, without ever consulting the counter directly. According to Wigner, it is only at this final stage of the proceedings that reality appears, because reality owes its creation to the conscious act of observation by the experimenter, or anyone else. Thus we must conclude that before the photograph was scrutinized, the nucleus, the Geiger counter and the photograph were all in schizophrenic states consisting of overlapping alternative outcomes to the experiment even though the delay before the photograph is developed could be many years. This little corner of the universe has to hang around in unreality until the experimenter (or a curious onlooker) deigns to glance at the photograph.

The real problem arises if two successive photographs, call them A and B, are taken of the Geiger counter at the end of the experiment. As the pointer is locked in place we know that the image on A must be the same as the image on B. The snag arises if there are two experimenters also, call them Alan and Brian, and Brian looks at photograph B before Alan looks at A. Now B was taken after A, but scrutinized first. Wigner's theory requires Brian to be the conscious individual responsible for creating reality here, because he looks at his photographic record first. Suppose Brian sees a deflected pointer and pronounces that the nucleus had decayed. Naturally, when Alan looks at photograph A it will likewise show a deflected pointer. The difficulty is that when photograph A was taken, B did not even exist, so in some mysterious way Brian's glance at B causes A to become identical to B even though A was taken *before* B! It seems that we are forced to believe in backwards causation; Brian looking at a photograph, perhaps many years later, influences the operation of the camera for the preceding photograph.

Few physicists are willing to invoke consciousness as an explanation for the transition of the world from a ghostly superposition to a concrete reality, yet von Neumann's chain has no other obvious end. We can consider larger and larger systems, each acting as a sort of observer of the other, recording the state of the smaller system, until the whole assembly encompasses the entire universe. What then? As we saw in chapter 5, the universe in fact should be described as a superspace of universes – a superposition of an infinity of overlapping worlds. If our world is just a projection out of superspace, or a three-dimensional slice through it, then some way has to be found of collapsing from the vast array of worlds in superspace onto this solitary projection. But as we now know, this collapse into reality requires an external non-quantum system to observe it. When we are dealing with the whole universe – all of creation – there is, by definition, nothing external that can observe it. The universe is supposed to be everything that there is, and if all is quantized, including spacetime, what can collapse the cosmos into reality without invoking consciousness?

One extraordinary idea which has enjoyed limited success among physicists was proposed by Hugh Everett in 1957 and developed by Bryce DeWitt of the University of Texas. The basic idea is to abandon the epistemological and metaphysical aspects of quantum theory, and take the mathematical description at face value. This is a subtle point which needs to be explained. When we use mathematics to model a familiar system, such as the path of a bullet, the progress of an economy or even counting sheep, the mathematical symbols are supposed to directly represent the things we are modelling (i.e. bullets, money or sheep). This remains true in much of modern physics also, and is certainly the case in Newtonian mechanics. In the conventional interpretation of quantum theory, however, it is not true. As explained in the foregoing chapters, it is necessary to describe the motion of a microscopic particle by a wave. The wave itself is not a physical thing that we can envisage as a substance or observe in the laboratory; it is a probability wave. Also, as the discussion of chapter 6 indicated, we cannot even regard the particle itself as a thing in its own right with independent qualities. It follows that the mathematics here refers to something quite abstract, and really only provides an algorithm for calculating the results of actual observations. According to Bohr, the matter wave is not supposed to be an objective thing at all, but merely a computational procedure. He claims that 'It is wrong to think that the task of physics is to find out how nature *is*. Physics concerns what we can say about nature.' And according to Heisenberg, mathematics 'no longer describes the behaviour of elementary particles, but only our knowledge of that behaviour'.

The proposal of Everett and DeWitt is to restore the reality of the wave and to regard it as a true description of the world. The pay-off for this promotion in status is the removal of the measurement paradox described above, because no special collapse into reality need occur at the moment of observation – the reality is already there. Thus, in the Everett theory, we can regard subatomic particles as really existing in a definite, well-defined condition, though they are still subject to the usual quantum mechanical uncertainties. This is in marked contrast to the Copenhagen interpretation described in chapter 6.

In view of the discussion given in the previous chapter about the difficulties surrounding a commonsense view of reality, it might seem strange that a simple change of perspective regarding the mathematics should restore reality. The point is that the Everett picture of reality is as far removed from the commonsense one as the Copenhagen picture. The ability of waves to overlap and of quantum states to build up from a superposition of other states is an inescapable component of microscopic physics. In the Everett theory this is serenely accepted and carried to its logical conclusion: if the wavelike superposition is real, so is superspace. Instead of assuming that all the other worlds in superspace are merely potential realities – failed worlds – that jostle our world of experience but do not themselves acquire concreteness, Everett proposes that these other universes actually exist and are every bit as real as the one we inhabit. Indeed, as we shall see, it is wrong to think of us as inhabiting one particular world of superspace: in the Everett theory, superspace itself is our home.

The Everett theory is sometimes called, for obvious reasons, the many-universes interpretation of quantum theory, and it has some remarkable implications, one of which is well illustrated by the polarizer and photon. As explained in the previous chapter, if a polarizer is placed at a particular angle, a photon will either be passed – in which case it emerges with precisely the polarization angle of the polarizer – or it is blocked. In terms of waves, the state of the photon before it reaches the polarizer is a superposition of two worlds, one in which the photon's polarization is lined up parallel to the polarizer, and the other in which it is perpendicular. Now the Copenhagen interpretation says that, on reaching the polarizer, only one of these two worlds is projected out of superspace as the true reality. In the many-universes theory, both are real, so the act of shooting a photon at a polarizer literally splits the world into two: one with a passed photon, the other with a blocked one.

In the above discussion, a particularly simple example was chosen in which only two alternatives were available. In general, however, there will be many more alternative worlds available for the outcome of an experiment, and there may even be an infinity of them. It follows that, according to this theory, the world is continually splitting into countless near copies of itself. In the words of DeWitt 'Our universe must be viewed as constantly splitting into a stupendous number of branches'. Every subatomic process has the power to multiply the world, maybe an enormous number of times. DeWitt explains: 'Every quantum transition taking place on every star, in every galaxy, in every remote corner of the universe is splitting our local world into myriads of copies of itself. Here is schizophrenia with a vengeance!' In addition to this ceaseless replication, our own bodies are part of the world, and they too are split and split again. Not only our bodies, but our brains and, presumably, our consciousness is being repeatedly multiplied, each copy becoming a thinking, feeling human being inhabiting another universe much like the one we see around us.

The idea of one's own body and consciousness being split into billions upon billions of copies is somewhat startling to say the least, yet the proponents of this theory have argued that the splitting process is quite unobservable, because the replicated consciousness cannot communicate in any way with its siblings. In fact, the separate worlds of superspace are all completely disconnected from each other as far as communication is concerned. It is not possible for an individual to leave one world and visit his copy in another, nor can we even glimpse what life is like in all those other worlds.

If we cannot see all these other worlds, or visit them, where are they? Science fiction authors have often invented 'parallel' worlds that are supposed to co-exist 'alongside' ours or to 'interpenetrate' ours in some way. In a sense, many people have an image of heaven as an alternative world co-existing with ours but not occupying the same physical space or time. Attempts are sometimes made to explain ghosts as alleged images from some other world briefly glimpsed by people endowed with special sensory abilities. As far as the scientist is concerned, our world is experienced as four-dimensional (three of space, one of time) but frequently additional dimensions are grafted on, either for mathematical convenience or, as in the case of Everett's superspace, as a model of reality. Mathematically, these extra dimensions are easily handled, although they may be hard to visualize physically. Ironically, rather than being parallel to our space, any extra dimensions of which we are not aware are described mathematically as perpendicular to ours.

To understand this point, imagine the experiences of a totally flat creature – let us call him a pancake – living on a two-dimensional sheet, such as the surface of a table or a ball. For the pancake his whole world consists of this two-dimensional surface, and he cannot perceive 'up' or 'down'. Things in the pancake's world have extension described by length and area, but he has no concept of volume. With our superior perception, we can see that the pancake is really embedded in a larger space that extends away perpendicular to the pancake and his surface. We can see that there is an outside and an inside of the ball, an idea which the pancake could be taught to understand and describe using mathematics, but one which he would have difficulty in visualizing in terms of familiar physical concepts.

Similarly, if there existed extra directions in space perpendicular to height, length and breadth, the restriction of our perceptions to these three dimensions would prevent direct knowledge of them, though we might infer their existence using mathematics and experiment. In Everett's world model, space is just one three-dimensional subspace from a superspace that really contains an infinity of perpendicular directions, an idea totally impossible to visualize, but with a sound basis in mathematics.

Although we cannot perceive all these other worlds, their existence leads in a very natural way to the statistical properties of quantum systems, which in the usual interpretation of quantum theory emerges as an inherent element of nature without explanation. As explained on page 112, we normally use the concepts of statistics and probability when we lack detailed information about a system. For example, when tossing a coin, because we do not know in detail the rate of spin, height of flip, etc. we can only say that there is a fifty-fifty chance of heads, or tails. Thus, the uncertainty is really just a measure of our ignorance. In quantum theory the uncertainty is absolute, for even the most detailed knowledge of the state of, say, a radioactive atomic nucleus, fails to predict exactly when it will decay. The many-universes theory gives a new perspective to this fundamental indeterminacy. The information which would have led to complete predictability is, crudely speaking, hidden from us in the other worlds to which we have no access. Thus, superspace as a whole is completely deterministic; the random element comes from our sampling just a minute portion of the whole. Regarding the real universe as the whole of superspace, one sees that God does not, after all, play dice. The game of chance comes not from nature, but our perception of it. Our consciousness weaves a route at random along the ever-branching evolutionary pathways of the cosmos, so it is we, rather than God, who are playing dice.

Many of the other worlds are very similar to our own, differing only in the condition of a few atoms. These contain conscious individuals virtually indistinguishable from ourselves in body and mind, acting out almost parallel existences. Indeed, these near duplicates share with us common precursors, for towards the past the branches converge and fuse. So what starts out at birth as one consciousness multiplies countless billionfolds by death.

Not all the other worlds are inhabited by our other selves, though. In some of them, the branching paths lead away to premature death. In still more, no birth will have occurred, while others may have deviated so far from the world of our experience that no life of any kind is possible. This topic will be covered in the next chapter.

What can we say about the other regions of superspace of which we are but a tiny sample? What is going on in all those other worlds? In chapter 1 we decided that some processes, like throwing a ball, are relatively insensitive to small changes in initial conditions whereas others, like the motion of a collection of snooker balls can be drastically altered by the slightest variation of speed or angle of the cue ball. In superspace, the quantum indeterminacy will cause balls, and everything else, to follow slightly uncertain paths. Each world of superspace is a separate reality with its own path for the ball, so that each point in Figure 2 represents a genuine universe, slightly different from its neighbours. In many cases, where small disturbances do not make much qualitative difference, the worlds will be almost indistinguishable, but where the process concerned is delicately balanced on the scales of chance, the alternative worlds will differ markedly.

One important example of how quantum phenomena can influence the world of our experience in a drastic way concerns the effect of radiation on genetic material. The make-up of all terrestrial living material is controlled by the long chain molecule known as DNA, which is a double helix of atoms arranged in a complex pattern. If the pattern is altered in any way, the genetic code is changed and the DNA will not correctly reproduce. If the altered DNA is an egg or sperm cell, the offspring will be a mutant. DNA can become damaged in many ways, but a universal threat is that of cosmic radiation – high energy subatomic particles which pepper the Earth from outer space. The impact of an energetic particle with a DNA molecule can result in a mutilated genetic code.

Mutations are vital to evolution, because they supply a variety of alternative life-forms for nature to select or destroy according to their efficiency. But as far as an individual person is concerned, mutation can be disaster. Evidently the occurrence of a mutation is an exceedingly delicate matter, for it depends on a subatomic particle colliding with a certain part of a molecule. The particle itself might well have been created as a secondary, high up in the atmosphere, when a primary particle smashed into the atoms of air. It follows that even an infinitesimal change in the angle of flight of the secondary would be sufficient for it to miss the precise molecule miles below, and the mutation would not occur. So we see that genetic accidents are exceedingly unstable to small subatomic changes, and the neighbouring worlds of superspace could be very different as far as a mutant person is concerned. Moreover, if the mutation resulted in some superior quality – such as great literary, military or scientific ability – then the world inhabited by the mutant could be drastically changed by his influence. Alternatively, historically vital figures will, in neighbouring worlds, have suffered deleterious mutations and will not rise to prominence.

By going back far in time, very small changes can produce large differences now. For example, in a world where an accident befell one of our ancestors ten thousand years ago, all his descendants living now, which could amount to thousands of people, would not exist. To take another example, exceedingly slight changes in the motions of the planets or the rocky debris between them could change a harmless near miss of an asteroid into a horrendous cataclysm.

Taking the widest possible view of superspace, it seems that every situation that can be reached along some convoluted path of development will occur in at least one of these other worlds. Every atom is offered billions of trajectories by the quantum randomization, and in the many-worlds theory it accepts them all, so every conceivable atomic arrangement will come about somewhere. There will be worlds that have no Earth, no sun, even no Milky Way. Others may differ so much from ours that no stars or galaxies of any kind exist. Some universes will be all darkness and chaos, with black holes roaming about swallowing up haphazardly strewn material, while others will be seared with radiation.

Universes will exist that look superficially like ours but have different stars and planets. Even those with essentially the same astronomical arrangement will have very different life forms: in many, there will be no life on Earth, but in others life will have progressed more rapidly and there will be Utopian societies. Still others will have suffered total destruction from war, while in some the whole Milky Way will be colonized by aliens, Earth included. In fact, there is virtually no limit to the availability of alternatives.

This vast multiplicity of realities raises an intriguing question: why do we find ourselves living in *this* particular universe rather than one of the myriad of others? Is there anything special about this one, or is our presence here simply random? Of course, in the Everett theory, we are living in many other universes also, but still only a tiny fraction are inhabited, for many could not support life. How many of the features around us are necessary for life to exist? These are the problems to be tackled in the next chapter.

8. *The anthropic principle*

Why is the world organized the way it is? The universe we inhabit is a very particular sort of place, full of elaborate structure and complex activity. Is there anything very special about the arrangement of matter and energy that we actually observe, in contrast to what might have been? Phrased differently, among the infinity of alternative worlds which surround us in superspace, why do our conscious minds perceive this particular one rather than another?

Questions of selection and probability must always be approached with caution. If a pack of cards is shuffled and dealt, the hand which each player receives is *a priori* overwhelmingly improbable; that is, if asked to predict one's hand before shuffling, the chances of success are exceedingly remote. Yet we do not, of course, regard each dealt hand as a miracle. Generally, one set of cards is much the same as another, and there is often nothing remarkable about some particular random selection. However, if we were dealt a complete suit we should regard that as an extremely strange occurrence indeed, for suit order has a significance above that of any other less structured sequence of cards. Similarly, winning a raffle is considered to be a lucky event, because the winning number, while no more remarkable than any other, has a special significance.

In the traditional religious approach to questions of the cosmic arrangement, it is usually supposed that the world was made by God with the particular structure that we encounter built in, precisely for the purpose of colonizing it with humans. The bible gives a direct account of how this was accomplished: first light was supplied, then a firmament in the midst of the waters; the waters were divided between those which were under the firmament and those above the firmament, and those under the firmament were gathered together into one place; dry land appeared and finally the Earth was stocked with plants and animals. In this way God created the conditions necessary for the support of human life.

An examination of life on Earth reveals just how delicately our existence is balanced on the scales of chance. There is a long list of indispensible prerequisites for the survival of our species. First there must be an abundant supply of the chemicals which make up the raw materials of our bodies: carbon, hydrogen, oxygen, as well as some small but vital quantities of heavier elements such as calcium and phosphorus. Secondly, there must be no risk of contamination by other chemicals which are poisonous: we would not want an atmosphere of methane or ammonia, as found on many other planets. Thirdly, we require a rather narrow range of temperatures so that our body chemistry can proceed at the correct pace. Without special clothing it is doubtful if humans could survive for long outside the temperature range 5°C-40°C. Fourthly, a supply of free energy is needed, which in our case is supplied by the sun. It is important that this energy supply remains stable and is not subject to large fluctuations, which not only requires that the sun continues to burn with extraordinary uniformity, but that the Earth's orbit be nearly circular to avoid excursions toward and away from the solar surface. A fifth requirement is that the Earth's gravity is strong enough to restrain the atmosphere from evaporating away into space, but weak enough so that we may move about easily and fall over occasionally without disastrous injury.

A closer inspection shows that the Earth is endowed with still more amazing 'conveniences'. Without the layer of ozone above the atmosphere, deadly ultraviolet radiation from the sun would destroy us, and in the absence of a magnetic field, cosmic subatomic particles would deluge the Earth's surface. Considering that the universe is full of violence and cataclysms, our own little corner of the cosmos enjoys a benign tranquility. To those who believe that God made the world for mankind, it must seem that all these conditions are in no way a random or haphazard arrangement of circumstances, but reflect a carefully prepared environment in which humans can live comfortably, a pre-ordained ecosystem into which life slots naturally and inevitably – a tailor-made world.

The status of these 'coincidences' changed dramatically when it was discovered that life on Earth is not static, but continually evolving. It then became possible, on the basis of Darwin's theory of evolution, to turn the problem upside down and ask, not why is the Earth so well-suited to life, but why is life so well adapted to the Earth. Mutation and natural selection supplied the answer: organisms that by random change find themselves slightly more attuned to the prevailing conditions have a selective advantage in the survival stakes, and will tend to proliferate at the expense of their less well adapted neighbours. Had gravity, for example, been stronger, it would have favoured the development of smaller, squatter creatures with stronger bones. A higher ambient temperature would encourage the development of cooling fins and other means of heat control. In many ways, therefore, there is nothing very special about the Earth after all, as far as life is concerned. Had conditions been different, so should have we.

Nevertheless it is not possible to argue that we could have evolved to fit any conditions whatever, because there are certain absolute limits and requirements without which no life at all is possible. For example, it is doubtful if life can exist on a planet with no atmosphere (such as the moon) or with a temperature well above that of boiling water. It is also hard to imagine life around a sun with erratic habits: many stars are known to flare up unpredictably and even explode. By appreciating that the sun is just a typical star we can view life on Earth in a more cosmic perspective. Stars come in all varieties of size, mass and temperature, and though our sun is a dwarf among stars, others of its type are not uncommon. There are so many billions of stars (maybe an infinity of them) that even if life is an incredibly rare accident it is clear that it will occur eventually in odd parts of the universe. That life has arisen on Earth is merely a consequence of the fact that the accident is most likely to occur on a planet whose conditions are optimal. We can conclude from this that our location in the cosmos is not a random one, but is selected by the conditions necessary for us to be here. This important conclusion, often taken for granted, could be vital to our view of ourselves and our place in the great scheme.

If the same reasoning is applied to our location in superspace as was just applied to our location in space, we can conclude that a great many other features of the world must be due to this biological selection effect. As only a small subset of all possible worlds can support life, most of superspace is uninhabited. The world we live in is, inevitably, the world we *live* in.

This type of reasoning is known, somewhat grandiosely, as 'the anthropic principle'. Its significance depends on which interpretation of quantum theory is adopted. According to the conventional Copenhagen interpretation, outlined in the earlier chapters, only our world 'really' exists, the other regions of superspace being 'failed' worlds – potential alternatives that nature, with random caprice, rejected. In which case we cannot say that our own existence explains the structure and organization of the universe (at least inasmuch as it affects the survival of intelligent life) because that would involve circular reasoning: we are here because the conditions are right, and the conditions are right because we are here. All that the anthropic principle can provide is a comment on how lucky it is that we are here. If a vastly greater number of alternative worlds cannot support intelligent life, then they would pass unwitnessed, with no cosmologists to wonder about how improbable they are. We should then have to regard ourselves as immensely lucky to be alive, and view our existence as an exceedingly improbable accident.

On the other hand, in the Everett many-universes interpretation of quantum theory, all the other worlds of superspace are real, each with an equal status of existence. If life is very delicate, then most of these worlds are even now devoid of observers. Only ours and those very similar thereto will have spectators. In this case we have, by our very presence, selected the type of world we inhabit from among an infinite variety of possibilities. Whether this is considered a true explanation of the world depends on the precise meaning to be attached to that word. If by explanation one means identifying the cause of something then, given the usual understanding of causality, we would not say the universe was actually caused by life, as life came afterwards. But if 'explanation' means a framework for understanding, then the many-universes theory does provide an explanation for why many things around us are the way they are. Just as we can explain why we are living on a planet near a stable star by pointing out that only in such locations can life form, so we can perhaps explain many of the more general features of the universe by this anthropic selection process. In short, the two interpretations of quantum theory amount to either chance or choice as an explanation of the world.

Just how delicately is life balanced on the scales of chance, and how widely could the features of our universe vary for it still to exist? Moreover, just how different are the other worlds of superspace? Could it be that nearly all of them, in spite of all the available variations, end up looking much the same as our own? To answer the first of these questions it is necessary to determine how large a fraction of all possible worlds are habitable. Right at the outset it must be re-emphasized that the nature of the world depends on two things: the laws of physics and the initial conditions. It was explained in chapter 1 how the shape of the path followed by a ball thrown into the air is determined (neglecting quantum effects) by both Newton's laws of motion as well as the angle and speed of projection. As the laws of physics are supposed to be absolute we should expect them to hold in the other worlds of superspace also. In contrast, the initial conditions that accompany any particular process will not be the same elsewhere, as that is precisely the difference between the separate worlds.

There are two difficulties about dividing up influences into initial conditions and laws of physics. The first is that in cosmology, where the subject of interest is the entire universe, it does not make much sense to talk about a law of physics. A law is characterized by the property that it applies repeatedly and unfailingly to a large number of identical systems, but as there is only one universe accessible to our observation we cannot verify whether it behaves (as a whole) according to some law. For example, is it a law or merely an incidental feature that the temperature of space (well away from stars) is about three degrees absolute: could it have been some other temperature? Only if we could see the other worlds of superspace and check that our suspected lawlike features were displayed there also, could any laws of cosmology be established. The second difficulty is that what might appear to one generation to be a fundamental law of physics may turn out in a subsequent generation with superior scientific knowledge to be merely a special case of some still more fundamental law. One familiar example concerns night and day. To the ancients it was a law of nature, equal in status to all the others, that the day is unfailingly twenty-four hours in duration. With our superior knowledge of mechanics we now know that there is nothing fundamental about the period of twenty-four hours, and that the length of the day can, and does, vary. The variations are only slight (though easily measurable with modern atomic clocks) over a human lifetime, but over geological time scales the length of the day has increased by several hours. When it comes to considering the other worlds of superspace we have to decide which features of our world are candidates for variation, i.e. which are the incidental features, like the length of the day on Earth, and which are really basic. As we do not know which of our laws are only special cases of more general laws, the safest strategy is to first consider variations of the things that are known to be incidental, then allow the presently accepted laws to vary, bearing in mind the speculative nature of the analysis.

The sort of question we should like to answer is whether we could live in a universe where the temperature of space is three hundred rather than three degrees. To answer such a question requires a definite idea of what is meant by 'we'. If 'we' means intelligent life of the form encountered on Earth the answer is probably no: it would be too hot for terrestrial life anywhere in the universe. On the other hand, there may be forms of life quite unlike terrestrial bioforms, perhaps based on entirely different processes, which could survive and even flourish under widely different conditions from those on Earth. Terrestrial life is based on the element carbon, which has the important chemical property of being able to link its atoms together, and with other atoms such as hydrogen and oxygen, in an enormous variety of ways. The key to life is complexity, for without an enormous number of possible variations among living organisms, evolution could not proceed. Life must be able to adapt in an almost limitless number of ways to the prevailing conditions and, as explained, this it does by occasionally producing random errors in the chemical make-up of one individual. After a huge number of useless errors, a species will hit upon a slight variation that endows the individual organism with some feature that still better suits it to the environment at the time. In this way, proceeding in billions of tiny steps, intelligence has evolved on Earth.

The need for sufficient complexity greatly limits the available chemical elements on which life can be based: perhaps carbon is the only one, though silicon and tin are occasionally proposed as possible candidates. The problem is that there is no real definition of life. Living systems are examples of organized matter and energy at extreme levels of complexity, but no boundary exists between the living and non-living. Crystals, for example, are highly ordered structures which can reproduce themselves, yet we do not regard them as living. Stars are complex and elaborately organized systems, but are not normally thought of as alive. It could be that we are too narrow-minded in our vision of life: there may be complex systems in other regions of the universe that bear no resemblance at all to the living organisms that are present on Earth, and yet are every bit as living as we are. One speculation about a bizarre alien life form was made by the astronomer Fred Hoyle in his science fiction novel *The Black Cloud*. The subject of

Hoyle's conjecture is the huge clouds of gas, mainly hydrogen, which roam about interstellar space. These gas clouds are nothing like clouds on Earth, and by terrestrial standards are exceedingly tenuous, containing only about one thousand atoms per cubic centimetre, which is a billion-billionth of the density of air and would therefore be considered an excellent vacuum in the laboratory. Nevertheless in the near-perfect emptiness of space the clouds are very substantial bodies and scatter a great deal of light. In the novel, Hoyle claims that some of these clouds are actually living, in the sense that they are motivated, and control their motion much like an amoeba; they possess a complex internal organization, including intellectual capacities far in excess of humans.

All forms of chemical life are essentially electromagnetic in nature; that is, the forces which control the chemical processes in our bodies are the electric and magnetic forces which act between atoms. But electromagnetism is only one of four known forces of nature. In addition there is gravity, and two nuclear forces, known as weak and strong. It is important not to exclude the possibility of life based on these other forces in any general assessment of the conditions necessary for life to form. However, superficially at least, the other three forces do not appear to be a realistic basis for life. Gravity is so weak that only astronomical masses exert significant forces. A galaxy, or at best a cluster of stars, seems to be the only type of gravitationally organized system known. Can a galaxy be in some sense living? It is hard to know how this can be the case. Aside from all else, it takes light, the swiftest entity, tens of thousands of years to cross a galaxy, which means that, according to the theory of relativity, a galaxy can only execute integrated behaviour patterns on this time scale. Put differently, the 'thinking time' for the Milky Way is about 100,000 years, so any organized activity must be slower than this, which is sluggish by any standards.

Nuclear forces have their problems too. Atomic nuclei are composite bodies bound by the strong force, so at first sight resemble molecules in which atoms are bound by electromagnetic forces. The resemblance is only slight, though. Nuclei consist of two types of particles: one called protons, which are electrically charged, and the other called neutrons, which are not. Both types experience a strong nuclear attraction which causes them to stick together in a blob. A heavy nucleus such as that of the uranium atom has over two hundred particles cohering in this way. The reason that nuclear life seems impossible is to be found in the balance of forces inside the nuclei. The strong nuclear force tries to glue the particles together, but the electric force on the protons is a counteracting disruptive influence, because each proton repels all the others electrically at the same time as it attracts them with the nuclear force. Although the nuclear attraction is much stronger than the electric repulsion, it is very short in range, and drops to almost nothing when the particles are separated by more than about one ten-million-millionth of a centimetre. This means that a neutron or proton will only attract its nearest neighbours, whereas the repulsion between protons operates on all the protons in the nucleus, because its action diminishes only gradually with distance. The disparity in range thus favours electric repulsion over nuclear attraction in nuclei containing many protons.

If the total electric repulsion grows strong enough, it can overwhelm the gluing force of the nuclear attraction, and the nucleus will explode. To help bring the forces back into balance, a heavy nucleus containing many protons can enlist the help of the neutrons, because they can assist in the gluing process without contributing to electric repulsion, as they are electrically neutral. So it is that light nuclei usually contain equal numbers of protons and neutrons (for example, oxygen contains eight of each) but uranium, the heaviest element found naturally on Earth, has ninety-two protons and up to 150 neutrons. Nuclei with still more protons are known but, like uranium, they are all radioactive, and disintegrate spontaneously. Evidently there is a limit to the number of neutrons that can bale a nucleus out of trouble, and the origin of this further instability concerns the other type of nuclear force, the so-called weak force.

The weak force is much weaker than electromagnetism, and its range is so short that it has never been measured as a finite extension. It does not play a role in binding particles together; its activity seems restricted instead to causing subatomic particles to scatter or disintegrate. The most dramatic example concerns, in fact, the neutron. If a neutron is freed from a nucleus, after about fifteen minutes it explodes into a proton, an electron and another type of particle called a neutrino. This speedy demise is prevented within the confines of the nucleus because of a fundamental quantum principle known as the Pauli exclusion principle, already discussed in chapter 4, which says that as all protons are identical, no two protons can (roughly speaking) occupy the same quantum state. That is, the waves of two protons must not overlap too much, which in physical terms means they can't approach too closely. Thus, if a neutron tries to decay into a proton, there may be nowhere for the proton to go if all the available sites in the nucleus are already occupied. Its decay would therefore be inhibited.

The structure of the nucleus is similar in one respect to that of the atom: atomic electrons are confined to definite energy levels, and both the protons and neutrons are also confined to energy levels within the nucleus. When the lower levels are full, an additional particle can only be accommodated in the nucleus by occupying one of the higher energy levels. In most nuclei, the neutron does not have enough energy to place the proton in one of these high energy levels, but if a nucleus acquires a great many neutrons then this problem is overcome. The reason is that neutrons are also subject to the Pauli principle, so the excess ones must find high-energy locations. In this elevated position they will have enough energy so that on decay the proton will be left in an unoccupied, high-energy site. It follows that neutron-rich nuclei are unstable and will convert spontaneously into nuclei with more protons, whereas proton-rich nuclei are electrically unstable and tend to fall apart. These two types of instability lead to two types of radioactivity, known as beta and alpha, respectively. Between them they ensure that no nuclei with more than a couple of hundred particles can exist for very long. This is nowhere near the level of variety and complexity needed for living matter.

In conclusion, it seems that the electromagnetic force is the only one capable of producing sufficiently complex composite bodies that meet any sensible definition of living. We arrive, then, at a definition of life as organized electromagnetic energy, probably based on chemical bonding. In what follows we shall adopt a conservative approach and assume that the only sort of life that can exist is similar to that encountered on Earth.

Turning now to the conditions in the other worlds of superspace and their suitability for life, it is first necessary to place the discussion in a cosmic perspective. We are not interested in other universes where life is not found on Earth, but occurs elsewhere nevertheless; our main concern is whether life can form anywhere in a particular alternative universe. According to our present understanding of astronomy, the sun is a typical star, so we should expect on general grounds that other similar stars will be associated with planetary bodies like the solar system. Planets are too small to be seen in even the most powerful telescopes, so we only have indirect evidence for their existence in other star systems. In spite of this, from what is known of the way that stars form, and from the existence of miniature versions of the solar system around Jupiter and Saturn (both have several moons), it is thought likely that most stars have planets, some of them inevitably similar to Earth. Our galaxy, the Milky Way, contains about one hundred billion stars grouped together in a gigantic spiral assembly typical of the billions of other galaxies scattered throughout the universe. This means that there is nothing very special about the Earth, so probably life is not a remarkable phenomenon either. Although we have no supportive evidence at all, it would be surprising if life were not widespread throughout the cosmos, though it may be rather sparse. The number of stars is so large that even if life is exceedingly improbable, it is still likely to have occurred somewhere else. If other universes exist in which life cannot form, it will be because the overall conditions have not come out right and the large scale structure of that universe is very different from our own. The condition of the Earth or the sun is far too parochial a concern to be of significance within the context of the anthropic principle.

Given that it is the large scale structure of the universe – the subject known as cosmology – that is of concern, we need not worry too much about the other worlds in superspace which are branching away from our own just now, for these will mostly resemble ours quite closely in their gross features. The reason for this is that the slight relocation or change in motion of individual atoms might, as suggested be responsible for altering the genetic make-up of a future leading politician, thereby causing or averting world war, but it cannot reshape the entire galaxy.

If we want to examine the branches that lead to substantially different worlds we have to trace them back to a common trunk. The greater the difference, the farther back we must go. The situation is similar to the random evolutionary changes of living things. Life began on Earth between three and four billion years ago with a few simple organisms, and from these common precursors new types gradually evolved. With increasing complexity the variety of forms increased also, until now we see living things as different as elephants, ants, bacteria and trees. Each generation witnesses new types branching away from the main species, but the steps are small and the process is very slow, so there is very little difference in a few generations. Consequently to trace back, say, monkeys and men, or sheep and goats, to a common origin, we only need to go back a few million years. To find the common trunk that branched one way to man and the other to mice, we must go back two hundred million years. A time twice as long as that is needed to find the common ancestors of man and frog, and still earlier epochs must be examined before animals and plants converge.

Tracing the branches of superspace back to a common origin is like seeking the origin of life on Earth. As explained in chapters 2 and 5 modern cosmologists believe the universe also had an origin, about fifteen billion years ago. On page 102 it was mentioned that the origin could have been a so-called spacetime singularity marking the ultimate past extremity to the physical universe. If this is correct, the singularity had no past of which we can know. In the moments after the singularity was the famous big bang, a primeval phase in which the expansion of the universe was explosively fast. To examine the fate of the other branches of superspace we must go back to this big bang and see how the alternative worlds emerged from the cosmic maelstrom. Just as small changes in organisms on Earth three billion years ago have led to greatly different evolutionary branches today, so random changes in the primeval universe could have set the world on a course leading to a condition today that would be unrecognizable to us. The cumulative effect of countless small changes drives the worlds of superspace on ever more divergent paths.

The changes of real interest to us concern the geometry of space. In chapter 5 the idea of superspace as a space of spaces was introduced. We can envisage each world as having a different geometry, in some the merest distortion, in others differences so great that even the topology is changed. Among the countless worlds of superspace there exist somewhere those universes with all conceivable shapes. The question of interest is whether our observed universe has a shape that is in any way special or remarkable and if so what relevance that fact might have to the existence of life in our universe.

The notion of the shape of space is a little vague, and some way has to be found to formulate the problem in a precise mathematical language. Mathematicians have invented quantities that provide a measure of the deviation of space from flatness, which is to say they gauge the distortions – bumps, twists, shears – at each place. Two types of distortions are easily recognized. The first is called anisotropy, and is a measure of how much the shape, or geometry, of space varies in different directions. For example, if along a certain line of sight the universe were very stretched out and expanding fast, whereas along a perpendicular direction it was shrunken and expanding slowly (or even contracting) we should say that the universe was highly anisotropic. The other type of distortion is called inhomogeneity, and this is a measure of how the geometry varies from place to place. If space contains a lot of irregularities and bumps, and is expanding at very different rates in different regions, it is very inhomogeneous.

It is obvious by glancing at the sky that the universe is neither exactly isotropic, nor exactly homogeneous. The presence of the sun, for example, causes a bump in space which represents a local inhomogeneity. The Milky Way defines a special direction in the sky, representing some anisotropy, this time of not so local an origin. However, when the really big telescopes are turned on extra-galactic space a remarkable thing is discovered. On a very large scale – that is, over distances about the size of a large cluster of galaxies – space seems to be highly uniform, both isotropic and homogeneous. In whichever direction astronomers look, they see roughly the same number of galaxies out to any particular distance, and furthermore, those galaxies at any given distance appear to be receding from Earth at about the same speed. Evidence for homogeneity is less good, but there is a geometrical connection between homogeneity and isotropy, which is this. Unless the Earth just happens to be located at the centre of the universe, which is a privileged status that is unthinkable these days, then if the cosmos appears isotropic about us here it must also be isotropic elsewhere. But a universe which is everywhere isotropic can be shown to be homogeneous also. It follows that either we are at the centre of the universe, or it is homogeneous as well as isotropic – at least on the very large scales under discussion.

If the universe really is homogeneous everywhere (and not just out as far as our instruments can probe) then it implies that there can be no centre or edge, as these would represent special locations, contradicting the homogeneity assumption. As explained in chapter 5, this does not necessarily mean the universe is infinite in extent, for space could curve around and join up with itself in a sort of hypersphere. These are issues of topology rather than geometry, and are probably not especially relevant to the anthropic principle and the conditions necessary for life, although they are of great interest to cosmologists and philosophers for other reasons.

In view of the unlimited variety of complex shapes that the universe could assume, it is a surprise indeed that the universe we observe turns out to have such a highly symmetric structure. So remarkable is this uniformity that most cosmologists are not prepared to accept the fact without inquiring how it has come about. It is all the more amazing when the theory of relativity is taken into account. We know that the speed of light plays a central role in this theory, inasmuch as no physical influence can propagate faster than light. When space is expanding, the behaviour of light can be rather strange. Just as a runner on a moving track has difficulty in maintaining headway, so when light spreads out into the expanding universe it is chasing after receding galaxies. The galaxies recede from each other because the intervening space is steadily stretching in all directions, so the space through which the light travels is stretching all the while along the line of flight of the light ray. One effect of this expansion is to stretch the light ray also, which increases its wavelength, giving it a redder colour. This is the origin of the famous cosmological red shift which Hubble first detected in the 1920s, and which he used to deduce that the universe was expanding.

As the light ray travels on, its wavelength gets longer and longer, and the question arises as to whether it will eventually be stretched by an unlimited amount, i.e. to an infinite wavelength. If that is so then it will be unable to convey any information at all. A mathematical analysis can reveal the circumstances under which this will happen. It turns out to depend on the precise way in which the universe expands away from the singularity. If it expands at a uniform rate, i.e. always doubling in size in the same interval of time, then light can always reach a distant place without being redshifted into oblivion. On the other hand, if the rate of expansion is not constant, infinite wavelengths occur. In particular, if the expansion rate slows down with time, then around any place in the universe is a sort of invisible bubble, representing the region of space which can be seen by an observer. The region outside the bubble cannot be seen, however powerful the instruments available, because no light from that region reaches the observer due to the infinite wavelength shift. The surface of the bubble thus plays the role of a sort of horizon, beyond which one cannot see. The bubble is centred on a particular observer: neighbouring observers have overlapping bubbles, but an observer on, say, the Andromeda galaxy (a galactic neighbour of the Milky Way) could see something on the edge of the visible universe that is inaccessible to us, and vice versa. When the observers are very far apart, their bubbles do not overlap, and to all extents and purposes they lie in physically different universes.

To test whether there is a horizon in the real universe, we can use mathematics. Einstein's general theory of relativity supplies an equation which connects the motion of space with the material content of the space, i.e. the gravitating matter. Solving his equation for the simplified case of a uniform universe leads to the result that so long as the energy and pressure of the matter remain positive (no counter-examples are known) then the expansion motion must decelerate. The explosive expansion of the big bang has slowed down progressively. Nowadays it is nearly a billion billion times slower than it was when the universe was about one second old. The conclusion is that there does indeed exist a horizon in our universe.

The bubbles do not remain static – their surfaces expand with the speed of light – which means that as time passes, more and more regions of the universe come into view. Crudely speaking, then, the horizon grows with the speed of light. It follows that the distance to the horizon must be the distance that light has been able to travel from the centre of the bubble during the age of the universe. At this time, therefore, the distance to our horizon is about fifteen billion light years. If we could see right to the edge, we should be witnessing the birth of the universe. Unfortunately, until about 100,000 years after the big bang, the universe was opaque to light, so the farthest back one can see is to that epoch. Information about earlier epochs comes from indirect sources.

The relevance of the horizon to the nature of the cosmological expansion can be understood by considering earlier and earlier moments, going back towards the singularity and origin of the universe. At one second after the big bang, the horizon was only one light second across – which is 186,000 miles. At one nanosecond, it measured scarcely more than a foot, and at the earliest time that we can sensibly mention, i.e. one jiffy, the horizon encompassed a volume of space so small that the number of 'bubbles' that would fit into a thimble is one followed by one hundred zeros. Now the bubbles represent regions of space which can have no physical communication whatever with other regions of space outside: the surface of the bubble is the greatest distance that the centre of the bubble can know about. What goes on beyond this limit cannot physically affect what goes on inside the bubble. Turning the clock back to one jiffy we are at the point where quantum fluctuations disrupt spacetime so severely that it ceases to exist as a continuum at all and behaves more like a foam. Inside the jiffy even the Everett branching has little meaning, so we can regard the jiffy as the starting point in the great cosmic drama.

What shapes of space emerge from Jiffyland, where all types of geometry exist in overlapping wavelike superpositions? Because the horizon is so shrunken at this point, every wormhole, every bridge, every tunnel in the foam of Jiffyland is comparable in size to the horizon, so the pattern of expansion initially reflects the particular local chaos of the quantum era. However, on a larger scale, the shape of space can be anything at all. As the different bubbles can know nothing of each other there seems to be no reason why they should all expand at the same rate.

It is here that we run into one of the great mysteries of cosmology. As explained, observations show that the universe is highly symmetric and uniform, both in the way that the galaxies are distributed through space and in the pattern of expansion motion. If the universe that erupted from Jiffyland consisted of myriads of causally independent regions, why should they all have cooperated to form a smooth and orderly motion? If the universe began at random, it would have started out by expanding in a highly turbulent and chaotic way, with each bubble of space, enclosed in its own private world by its horizon, exploding differently. No physical influence connects the bubbles together so they had no reason to cooperate. If the energy is distributed at random among all possible modes of expansion, most of the energy would end up in the chaotic motions, and only an infinitesimal fraction would appear in the smooth, uniform, isotropic motion that we actually observe. Out of all the unremarkable chaotic motions with which the universe could have emerged from the big bang, why has it chosen such a disciplined and specialized pattern of expansion?

A helpful way of highlighting the remarkable nature of the cosmological expansion is in terms of the discussion given on page 26 about initial conditions. If we imagine plotting a diagram in which each point represents a particular initial expansion pattern for the universe, then only one point will represent an exactly homogeneous and isotropic expansion. Because we can only detect departures from uniformity above a certain minimum value, purely for technological reasons, all we can say is that the universe is very nearly homogeneous and isotropic, to a certain accuracy (about 0.1 per cent in the case of isotropy) so on our diagram there will be a small blob representing all possible initial conditions consistent with the high degree of uniformity we actually observe. Outside this blob are chaotic states. If the universe is actually chosen at random from all these possibilities it would be like sticking a pin into our diagram, and it is clear that the chance of stabbing the little blob is very small. Of course, the idea is pretty vague, because we do not know how to measure areas on the diagram, so the size of the blob is not well-defined, but the qualitative idea is sound enough: the likelihood of the present arrangement arising by chance appears to be negligible.

There is a helpful analogy to the expanding universe which should clarify the issue. Consider a large group of people in a tight huddle. Each person represents a region of space enclosed within its own horizon – a 'bubble' of space – so to represent the fact that there is no communication between bubbles we equip everyone with blindfolds. Thus, each person is ignorant of the behaviour of his fellows. The compact group represents the initial singularity, and when a whistle is blown, the people all start to run in straight lines away from the centre of the huddle: the universe expands. The group spreads out in a sort of ring. The runners have instructions that they must adjust their stride so that the ring remains as circular as possible as it expands, but none of the runners knows how fast his neighbours are running, so each picks a random speed. The result is, almost certainly, a ragged, distorted line, very far from circular. There is, of course, a small chance that purely by accident all the runners will match their strides, but it is obviously pretty unlikely. What is observed of the universe today corresponds to a ring of runners so nearly circular that there is no detectable distortion in its shape. How can this have happened: is it a miracle? About ten years ago an ingenious suggestion was made to try and explain this curious symmetry. In the language of the runners it amounts to the following. When the group explodes outwards, some runners will inevitably run faster than their neighbours. However, after a while, fatigue will set in and they will slow down. Their colleagues, on the other hand, will not have dissipated their energy so rapidly and will have enough stamina to catch up. The end result will be, after a long enough time, an approximately circular ring of rather exhausted runners, plodding doggedly outwards at a considerably reduced rate.

Translated into cosmological language, the idea is this. In the primeval universe, some regions of space expanded rather more energetically (i.e. faster) than others, and some directions were vigorously stretched while others spread out more sluggishly. Dissipative effects began to sap the energy of the more vigorous motions and slow them down, enabling the sluggish motions to catch up. In the end the turbulent and chaotic early state is damped down and reduced to a rather slow and quiescent motion, with a high degree of uniformity, precisely as observed.

For this explanation to work it is first necessary to find a dissipative mechanism, comparable to the runners' fatigue, that will erode the stamina of the expanding universe. This will have to come about in such a way that the energetic motions are discouraged more vigorously than the sluggish motions. Several candidate mechanisms exist. One possibility is ordinary viscosity – the effect which exerts drag on an aeroplane or boat. Another, which has been much investigated in recent years, is the spontaneous production of new subatomic particles out of empty space. This can happen because the energy of motion of space can be converted into matter according to the ideas of quantum theory and relativity outlined in chapter 4. Calculations show that particles of all types are produced by this mechanism – electrons, neutrinos, protons, neutrons, photons, mesons and even gravitons. The reaction on space caused by the appearance of all this new matter is to deplete the expansion energy and help to bring its motion back into line with neighbouring regions. A crucial feature of the mechanism is that its efficiency is greatest at the earliest moments when the expansion rate is so much faster. We would not, therefore, expect the primeval turbulence to survive very long; instead it would be converted into particles.

Whatever mechanisms are considered, the result of the dissipation of energy is ultimately heat. According to the universal second law of thermodynamics, which regulates the organization of all energy, any dissipative tendency inevitably generates heat. On Earth, the profligate dissipation of energy in our factories and homes produces so much heat that some scientists foresee the day when it will threaten the existence of the polar ice caps. In the primeval universe, the heat generation due to particle creation and other dissipative processes was colossal, and the big bang took on the features of a furnace, with temperatures far in excess of anything available in the universe now, even at the centres of the hottest stars. One of the most exciting scientific discoveries ever made occurred in 1965 when two American radio engineers accidently discovered the remnants of the primeval heat while working on satellite communication for the Bell Telephone Company. Because the universe is now enormously distended compared to the primeval epoch, this heat has cooled away to very nearly nothing, and only a remnant of the fiery birth of the cosmos remains at a temperature about three degrees above absolute zero. This cosmic background radiation coming from all directions of space apparently bathes the whole universe and is good evidence that the hot big bang theory is substantially correct. It also provides the best means available to test the isotropy of the early universe, for this heat radiation carries information from the time that the universe changed from opaque to transparent about 100,000 years after the beginning. At that time the temperature had fallen to a few thousand degrees and the primeval gases no longer absorbed the radiation. As far as we can tell, the universe at 100,000 years was isotropic to an accuracy of 0.1 per cent.

The primeval heat is also of crucial importance to our understanding of much earlier moments than 100,000 years. Very little is known about the detailed physics that controlled the cosmic material during the primeval phase between one jiffy and one second after the beginning: only a few basic principles and some mathematical analyses can help. We can, for example, try to compute just how much heat is created by dissipating a given amount of turbulence and comparing the answer with the observed three degrees, which then reveals how chaotic the primeval universe was. It turns out that the quantity of heat produced by a given amount of turbulence depends on the precise moment that it is converted. The reason for this is that the dissipation takes place while the universe is expanding and the expansion motion itself has the effect of reducing both the heat energy (which is why the primeval radiation is now so cool) and the turbulent energy. A mathematical investigation shows that the energy of turbulence falls much faster than the heat energy as a result of the expansion, which means that the earlier the conversion takes place from the former to the latter, the more heat energy we have at the end of the day. This simple piece of information raises a great paradox, because all the dissipation mechanisms, such as particle production, are at their most efficient early on. Putting in the numbers, one finds that almost any anisotropy whatever would have generated more heat than we actually observe. Indeed, we seem to have the minimum possible quantity of primeval heat in the universe.

It is not possible for the universe to produce no heat at all, for there must be *some* turbulence present in the primeval phase. This is because, at the end of one jiffy, quantum fluctuations of space would be present and these alone will cause irregularities. A crude calculation can show how much heat these basic quantum space fluctuations will produce and it comes out very close to the observed value. Evidently there has been little additional dissipation going on over and above that of the quantum turbulence.

Even if we were wrong about the dissipation mechanism there is another reason why excessive primeval turbulence seems unlikely. The contribution of the energy turbulence to the total mass-energy content of the universe can be calculated, and its effect on the overall expansion rate computed. It turns out that when the turbulent energy dominates it slows up the overall expansion rate appreciably. It is as though the universe, in churning about at random, neglects to get on with the general expansion. This delay produces an important secondary effect, in that the heat radiation inevitably generated by the quantum space fluctuations after one jiffy – the quantum heat – does not cool as rapidly as it would have done in a more vigorously expanding, smoother universe. The result is that we end up, once more, with too much heat. Either way, whether the turbulence dissipates directly to heat, or else slows up the cosmological expansion and prevents the quantum heat from cooling, the end result is to deliver more heat than we actually observe. It seems, therefore, that the cosmic background radiation is testimony to the fact that the universe was born in disciplined quiescence, right back as far as one ten-million-billion-billion-billion-billionth of a second – which is quite a conclusion!

If the above reasoning is correct, and some cosmologists are sceptical, it returns us to the paradox of why the universe began so smoothly. It is here that the anthropic principle may be of help. Although the primeval heat radiation is very inconspicuous – indeed, it needs special equipment to detect it at all – an increase in its temperature of just one hundred-fold would have drastic consequences for life. If the temperature exceeded 100°C then liquid water could not exist anywhere in the universe. Life on Earth would be completely impossible, and it is doubtful if life of any sort could form in the first place. An increase by a thousandfold would threaten the existence of the stars themselves, rivalling their surface temperatures and causing a build up of internal heat. Moreover, it is doubtful if stars or galaxies could have formed in the first place in the presence of so much disruptive radiation. From what we know of the dissipation of primeval anisotropy, it seems that even a tiny amount would increase the primeval heat billions of times. The temperature is therefore highly sensitive to any primeval turbulence. Nor does it make much difference that as the universe expands the temperature falls. At present it takes billions of years for the temperature to drop by one half, and all the stars will be burnt out by the time it falls to one per cent of its present value. If the formation of life has to wait that long, it loses the vital starlight on which it depends for energy.

Unless the connection between primeval turbulence and the cosmic heat radiation is totally misconceived, it comes as no surprise to find that the universe is expanding in such a smooth way. If it were not, we could not be here to wonder about it. We could regard our existence as an accident of almost unbelievable improbability – for out of all possible worlds, our universe happens to choose just that highly ordered arrangement of matter and energy that keeps the cosmos cool enough for life. Alternatively we could adopt the many-universes interpretation of superspace and say that among the countless numbers of turbulent, overheated worlds in superspace, there exist just a tiny fraction which are cool enough for life. The optimum conditions are to be found among the coolest, and that is where life most probably forms in abundance. It is no coincidence, therefore, that we find ourselves living in a world with a near-minimal primeval heat content. Most of the others are uninhabited. Of all the vast array of universes that exist, only in a minute fraction similar to our own are intelligent creatures pondering profound questions of cosmology and existence. The rest pass through their histories in roaring torment and searing heat, unnoticed by anyone – sterile, violent and apparently pointless.

9. Is the universe an accident?

In the previous chapter we discussed how an observer must encounter certain features in his world due to the fact that he can exist at all. If we believe in only one universe then the remarkably uniform arrangement of cosmic matter, and the consequent coolness of space, are almost miraculous, a conclusion which strongly resembles the traditional religious concept of a world which was purpose-built by God for subsequent habitation by mankind. On the other hand, if we accept the idea of an ensemble of many universes, such as the Everett interpretation of quantum theory suggests, then the structure of the universe is not an incredibly lucky accident, but a biological selection effect: we, as observers, have evolved only in those universes where the structure has this remarkable uniformity. In the many-universes theory, all of superspace is real, but only an infinitesimal subset is inhabited. The choice would seem to be philosophical rather than physical, and may merely reduce to a mode of speech. When one pools winner expresses gratitude to God while another proclaims his good fortune, are they really saying anything different?

In the last few years, the anthropic principle has been applied to a variety of other features of our universe on which life appears to be sensitively dependent. In addition to being highly isotropic, the universe on the large scale looks homogeneous – uniformly populated with matter. However, if it were too homogeneous, there would be no galaxies and, presumably, no life. The universe has therefore to strike just the right balance of clumpiness: too little and the cosmic matter remains as a disorganized gas. If, on the other hand, the material were more concentrated, there is a threat that it might disappear completely under the action of gravity.

Being a universal force, gravity attracts all matter to all other matter. The effect of gravity on a large ball of gas is to cause it to progressively shrink, and as it shrinks, gravitational energy is released and converted into heat, particularly near the middle. Eventually, as the interior temperature rises and the gas pressure builds up, the gas may be able to support the weight of the overlying layers: the shrinkage then stops. This is the situation with the sun and other stars, which are mainly in stable equilibrium at a constant radius. Of course, the heat cannot be retained in the ball for ever, as it tends to flow towards the surface and radiate away into space. If the lost heat cannot be replaced, gravity will take over once more and the shrinkage will continue. In stars, however, progressive shrinkage is postponed for a few billion years by an entirely different source of heat: nuclear burning.

Most of the matter in the universe is made of hydrogen, the lightest element that exists. Its atoms consist of just two subatomic particles, an electron and a proton, so the hydrogen nucleus is not a composite body as with other elements. Hydrogen is not the most stable material as far as nuclear structure is concerned. In chapter 8 it was explained how composite nuclei containing many protons and neutrons are bound together by a strong nuclear gluing force which overcomes the electric repulsion between the protons. In light nuclei, such as helium, oxygen, carbon and nitrogen, in which there are not many protons, there is a bonus from assembling the constituents together into a composite blob: the assembled nucleus is more stable than the separate particles. Energy is therefore released on assembly. Conversely, it would need a great deal of energy to overcome the attractive forces of the nuclear glue and pull these nuclei apart into individual protons and neutrons. In contrast, the heavy nuclei, such as lead, radium, uranium and plutonium, contain many protons, and there is actually an energy loss when more particles are added to the nucleus. This is because the combined electric repulsion of all the protons is greater than the attraction of the nuclear force, with the result that energy is released from the disintegration of heavy nuclei.

These facts are exploited in the nuclear power industry. The fission of heavy nuclei to release energy is the principle of nuclear power stations and the atom bomb, while the controlled fusion of light nuclei to release even more energy is still at the research stage. Uncontrolled fusion takes place in the hydrogen bomb, and also powers the sun and stars. In the sun's interior, hydrogen nuclei are being fused together into the next lightest element: helium. The helium nucleus contains two protons and also two neutrons, so during the nuclear burning two neutrons have to be obtained for each new helium nucleus. As discussed in chapter 8 a free neutron will decay into a proton after about fifteen minutes. What happens in the sun is the reverse of this process – protons change into neutrons, to assist in the synthesis of helium. The nuclear reactions which bring this about are complicated, but the net effect is to pass the discarded electric charge from the proton on to a positron (the antimatter image of the electron) which rapidly annihilates with a nearby electron to give gamma rays. Another byproduct of the process is a so-called neutrino, which immediately leaves the scene of the action and passes into space. The neutron combines with another neutron and two protons to form the nucleus of a helium atom, releasing more gamma rays in the process. After banging around inside the star for aeons, the gamma rays become converted into heat energy which helps support the star against gravitational shrinkage.

In all stars the nuclear burning will eventually cease when the fuel runs out, and the star will begin to shrink again. To find out what happens next we must use Einstein's general theory of relativity. A mathematical analysis shows that, so long as the star has less material than about three suns, other sources of pressure build up and the shrinkage can be arrested. For example, in some stars known as pulsars, the material is progressively crushed until even its atoms collapse and are smashed into neutrons. These neutron stars are balls of almost pure neutron matter and are incredibly dense, measuring only a few miles in diameter.

For stars with more than three solar masses their fate is still more bizarre. According to general relativity the shrinkage cannot be averted and they implode catastrophically, in about a microsecond. The escalating gravity in their vicinity so severely distorts spacetime that time there literally stands still. No light or matter, or any information, can escape from within this frozen surface, so it appears black – a black hole. The star itself, having retreated inside the black hole, effectively disappears from the universe. It may encounter a singularity inside the hole and hence leave spacetime altogether, but in any case, as far as the outside world is concerned, the material which made up the star has gone for good: nothing can return from inside a black hole.

Black holes are believed to play an important part in the final stages of our universe, when most stars will probably end their days inside one. However, they could also be important in the primeval stages too. The critical density of matter needed to form a black hole depends on the total mass. For the galaxy, the density of water is sufficient, but for the sun a density of a thousand billion kilograms per cubic centimetre is necessary. To form black holes smaller than one solar mass requires densities in excess even of this colossal amount. The only time such enormous densities have occurred in the universe was during the big bang, when the whole cosmos was exploding out of a condition of unlimited compaction. A number of cosmologists have investigated the formation of black holes in the primeval universe, but their results are rather inconclusive because they depend sensitively on the nature of the cosmic material at the enormous densities encountered there, and this is quite beyond our present understanding. However, on general grounds it is clear that if the material were very lumpy then black hole formation is more likely to occur than if the matter were smooth and evenly distributed. It seems safe to suppose that a universe that began in a very inhomogeneous condition would emerge from the big bang populated not by stars, but by black holes.

Could life form in a black hole universe? The black hole offers little prospect for a life-support system. Life on Earth depends crucially on the heat and light from the sun, and black holes, by their very nature, do not radiate energy of any sort (though, as will be explained shortly, this may not be true of microscopic black holes). In addition, instead of orbiting serenely around a star, a planetary body would, if it gets too close to a black hole, spiral inexorably inwards and rapidly plunge into oblivion down the hole itself.

How many primeval black holes are there? As yet nobody has definitely identified a black hole, though there are some strong candidates. The problem is, being black, they are hard to spot, and the only practical technique is to look for gravitational disturbances to more conspicuous bodies caused by their proximity to a hole. Individual black holes in the galaxy might show up by the effect they have on the motions of stars, while intergalactic supermassive holes might disturb the behaviour of whole galaxies. We could also assess the total mass of all the black holes in the universe by measuring the total gravity of the universe. This can be done by seeing how rapidly the expansion motion decelerates due to all the gravitating objects in the cosmos. Measurements indicate that the luminous matter (stars, gas, etc.) must form an appreciable fraction of the total mass of the universe, so it is clear that we do not inhabit a universe which is overwhelmingly dominated by black holes.

In spite of the absence of a detailed understanding about primeval black holes, it is possible to use a very general argument to estimate crudely how probable it is that the universe emerged from the big bang without an overwhelming number of them. Our ability to perform this calculation rests on some remarkable new results from mathematical work on quantum black holes – that is, the quantum theory of fields applied to black holes – initiated mainly by Stephen Hawking of Cambridge University. In 1974 Hawking showed that black holes are not really black at all, but emit heat radiation at a characteristic temperature that depends on their mass. This extraordinary result enables black holes to be treated rather like heat engines, and in particular it is possible to examine their properties by applying the universal laws of thermodynamics.

During the late nineteenth century, one of the great triumphs of theoretical physics was the discovery of a connection between the thermodynamic behaviour of a system and the probability of the atomic arrangements of its constituents. To give a simple example, imagine a box of gas: its molecules rush about at random, colliding with each other and the walls of the box. The pressure of the gas is caused by the combined impact of all the molecules, while the temperature is a measure of the molecular speeds. The heat energy is simply the energy of their motions. Thermodynamic quantities such as temperature, pressure and heat can be measured in the laboratory, but little can be known of the details of the individual molecules, for they are too small and too numerous to be perceived. Only the averaged, gross properties of billions upon billions of them are observable, so we do not notice their continual reshuffling and rearrangement as they collide with each other and move around. Any particular macroscopic state of the gas (i.e. temperature, pressure, etc.) could be produced by an enormous number of different internal arrangements of the molecules. For example, changing the positions of a few molecules would make no observable difference to the temperature.

Not all conceivable molecular arrangements lead to the same macroscopic state, however. For instance, in the unusual case that all molecules move to the left in unison, the gas would pile up in the left hand end of the box. If the molecules are all moving at random, why is this behaviour not occasionally encountered? The answer rests with elementary probability and statistics. The chances of such a cooperation occurring between vast numbers of individual molecules is incredibly small, though not necessarily zero. A far more probable pattern of motion is the chaotic one where the molecules are spread about more or less evenly, just as a jumbled sequence of cards from a shuffled pack is much more likely than an ordered suit sequence. The molecular collisions act as a random reshuffling mechanism, and the chances of them shuffling billions of particles into an orderly pattern of behaviour are negligible. This illustrates the very general principle that chaos is easier to achieve than order and hence is far more likely – reasoning which was applied in the last chapter to argue that an orderly primeval expansion of the universe was far less probable than a turbulent, chaotic state. But why is this so?

The reason why disorder is more likely than order can be found in the statistics of the molecular arrangements. As mentioned above small rearrangements of the individual molecules do not affect the gross properties of a gas. However, some states are more sensitive to rearrangements than others. For example, in the state where all the molecules rush the same way, there is not the same freedom to reshuffle the configuration as in a less orderly state, because even a minor disturbance is likely to break up the precisely coordinated behaviour. A mathematical treatment shows that the difference in rearrangeability between ordered and disordered states can be overwhelming. Some states – the highly disordered ones – can be disturbed in vastly more ways than all the other, more ordered, states. So if the molecular arrangements are being continually reshuffled at random, it won't take long for an orderly state to break up into a disorderly one, and once that is achieved the disorderly one is highly stable because subsequent reshuffling is more likely to reproduce another disorderly state than an orderly one. The principle is simplicity itself: there are vastly more ways of producing disorder than order, so a random state is overwhelmingly likely to be highly disordered.

Armed with the connection between the degree of disorder of a system and the likelihood of its state occurring from some random process, one may attempt to determine how black holes fit into this thermodynamic scheme, and assess the likelihood that they will appear as a result of purely random processes occurring in the primeval universe. At first sight the concept of the degree of disorder associated with a black hole seems somewhat obscure. Unlike a gas, which is known to be made up of billions of tiny molecules, a black hole is not really made up of anything at all, but is merely a vestige of vanished matter – a violently distorted region of empty space. Closer examination, however, reveals a deep similarity between the two systems. In both cases we do not have information about the internal structure: the gas molecules are too small for us to notice, while the interior of the black hole can communicate no information to the outside. All that can be measured in these systems are the gross features, such as the total mass, volume, electric charge, degree of rotation, etc. A particular set of values for these gross features can be achieved in an enormous variety of ways: the molecules of gas can be rearranged, while the same type of black hole can be made out of imploded stars with vastly different internal structures.

The really striking similarity between gases and black holes comes from the latter's compliance with a new law that seems to be a direct analogue of the central law of thermodynamics – the so-called second law. The second law states that total disorder always increases with time. The black hole obeys a law which says that it always gets bigger with time, so one suspects that the size of the hole is a measure of its degree of disorder. This guess was confirmed when the connection between the temperature of a black hole, as computed by Hawking, and its mass was examined: the black hole turns out to obey the same relation between disorder and temperature as does a gas, if the area of the hole is used as a measure of disorder. The area is in turn related to the mass of the hole so we have at hand a means of comparing the degree of disorder of a given mass of material with the equivalent disorder that would be achieved should that material be discarded down a black hole. In the case of one solar mass of matter, the black hole disorder works out at several billion billion times that of the actual sun, a result that carries an ominous implication: all else being equal, there is an overwhelmingly greater likelihood of the material of the sun being inside a black hole than in a star. The crucial qualification here is 'all else being equal'. Evidently all else is not equal in our universe, or there would be no sun, or any other stars. If the primeval material was churned about at random, it would have been overwhelmingly more probable for it to have produced black holes than stars, because the holes, being so much more disordered, can be produced in a vastly greater number of ways. For every star that formed, countless billions of more easily achieved black holes should have accompanied it.

The true force of these arguments is revealed when the exact mathematical relationship between disorder and probability is examined. It is, in fact, a so-called exponential relationship, similar to the way in which an ideal population grows, by doubling in size in some fixed interval of time, however large its size. Thus, every time the degree of disorder increases by some fixed amount, the probability of that state occurring doubles. The relationship is therefore an 'escalating' one, for when the numbers become large a small amount of additional disorder represents a hugely greater probability. In the case of the sun, whose disorder is only one hundred-billion-billionth of the equivalent black hole, the chances against the sun, rather than the hole, emerging from a purely random process will be roughly one followed by the same number of zeros! That is one followed by one hundred billion billion zeros, which is pretty improbable by any standards.

If the same argument is applied to the entire universe, the odds piling up against a starry cosmos become mind-boggling: one followed by a thousand billion billion billion zeros, at least. Even if the disorder-probability arguments have only very approximate validity, the conclusion must be that we live in a world of astronomical unlikelihood. Once again, the anthropic principle can be invoked to argue that among the overwhelming array of universes dominated by black holes, are an almost inconceivably tiny fraction in which, against all the odds, the primeval matter has escaped annihilation and arranged itself into life-supporting stars.

These considerations conjure up a bizarre spectacle of superspace: world upon world of chaotic motion, populated by huge black holes roaming about and colliding in titanic eruptions of spacetime, all bathed in searing heat generated from quantum noise and amplified by primeval dissipation. Who would imagine that among this infinite number of nightmare universes would exist an insignificant few which had miraculously manoeuvred away from the hole-ridden inferno and spawned life? We can imagine it, because we are that life.

As remarked at the beginning of this chapter, it seems that the universe must start out with some lumps and bumps in order for stars and galaxies to form in the first place. Although there is a natural tendency for gas clouds to shrink into blobs under the action of gravity, they have to fight against the expansion of the universe which has the opposite tendency, i.e. to disperse them. At one time astronomers hoped to account for the existence of galaxies by assuming that the material which erupted from the big bang was initially very smooth, but random fluctuations occurred which caused scattered accumulations of matter. These acted as nuclei around which other material settled due to the enhanced gravity there, so that gradually the gaseous material would have fragmented into distinct proto-galaxies, which in turn fragmented into stars. Unfortunately, it appears that too little time has been available since the beginning of the universe for galaxies to have grown naturally in this way. The only possibility is that some dense regions already existed at the outset, which subsequently became the galaxies that we now see. Thus, if the primeval material was too smooth, life would be just as impossible as if it were too lumpy.

So far, the discussion about the anthropic principle has been restricted to questions about the arrangement of matter and energy in the universe. It is possible to go beyond this and consider circumstances where the fundamental physical properties of matter can vary from world to world. As we found in chapter 8, we cannot know which of our laws of nature are merely special cases of more general laws, so many of the features of physics which we take for granted might be quite different in the other regions of superspace. To take a first example, our present theory of gravity (Einstein's general theory of relativity) has a built-in restriction that the strength of gravity between two standard masses a given distance apart is the same wherever in space they are located and whenever in time they exert their force. In the case of the Earth, it will pull on an apple with the same force whether the Earth is located in the Milky Way or the Andromeda nebula. Likewise, it will pull the apple equally hard today as it would have done a billion years ago. The law of constancy of gravity seems to be fairly well tested by experiment, though there is still room for doubt, and some physicists have proposed rival theories to Einstein's in which gravity can vary in strength from place to place and moment to moment. If the strength of gravity is not fixed once and for all by fundamental physical principles, one might suppose that it would vary from world to world in superspace. We then face a challenge to explain why it has the particular strength that it does in our universe; in particular, why it is much weaker than all the other forces of nature?

Anyone acquainted with elementary physics will know that the mathematical laws which describe fundamental physical systems frequently throw up numbers such as 4π or 12. Often these numbers have a geometrical origin, or are connected with the dimensionality of space. About fifty years ago, following the appearance of the general theory of relativity, many physicists attempted to construct a unified theory in which Einstein's gravity was combined with Maxwell's earlier theory of electromagnetism. The hope was, and still is, that in some way gravitational and electromagnetic phenomena may both be manifestations of a basic unified field. Nobody has succeeded in producing such a theory, although the search continues. One of the daunting difficulties facing unified field theorists is the gap in strength between the gravitational and electromagnetic forces. The gravity acting between the constituents of an atom is some forty powers of ten (10^{40}) less powerful than the electrical attraction. What theory of physics could possibly throw up such an enormously large number?

A curious twist to this mystery was first noted by the astronomer Eddington and the physicist Paul Dirac. When we measure intervals of time, we gauge them against some natural period of vibration or rotation: the rotation of the Earth, the oscillations of a quartz crystal or the undulations of a light wave. If we ask, what is the smallest natural unit of time which has fundamental significance for the structure of matter, we are led to examine the vibrations of atoms and their nuclei. The subatomic particles inside the nuclei of atoms oscillate on a time scale which is incredibly short by everyday standards – about a million-billion-billionth of a second, or the time taken for light to travel across a nuclear distance. This tiny interval of time forms a natural, fundamental unit against which to measure other intervals, though it is quite a thought that even this fleeting duration is twenty powers of ten longer than the natural unit of quantum gravity – the jiffy. Asking now what is the longest natural unit of time available, we are led to the age of the universe, which has been calculated in various ways to be about fifteen billion years. In our fundamental subatomic units this duration turns out to be about 10^{40}, or one followed by forty zeros – the *same* enormous number by which gravity is weaker than electromagnetism.

The mystery is, why do we happen to be alive at just the epoch at which the age of the universe is equal to the magic number 10^{40}? Dirac argued that this number is so much bigger than those normally encountered in physical theory, like 4π and 12, that it is most unlikely that the above two ratios are equal by coincidence. He maintained that the numbers are connected by a physical theory which requires the equality to remain true at all epochs, a feature which can be achieved by demanding that gravity grow weaker with time. In the far past, when the age of the universe was small, gravity was stronger than now.

Unfortunately there is little observational evidence for the weakening of gravity, and an alternative explanation for the 'coincidence' is suggested by the anthropic principle. The argument used here is adapted from that produced originally by the American astrophysicist Robert Dicke and the British mathematical physicist Brandon Carter.

It has already been remarked that the existence of heavy elements such as carbon is thought to be essential for life as we know it. Carbon was not present at the beginning of the universe (see below) but was synthesized in stars that died long before the sun was formed. It found its way to Earth because some of these stars exploded and spewed their carbon into interstellar space. It seems likely that life cannot form in abundance in the universe until at least one generation of stars has run its life cycle. On the other hand, once a star has died, perhaps to become a black hole or a cold compact object, it is highly improbable that life can form near it. As there are likely to be only a small number of successive generations of stars before most of the matter in the galaxies is burnt out, it follows that life will only arise in the universe in a period between one and a small number of typical stellar lifetimes.

Now the lifetime of a star can be estimated from the theory of stellar structure. It depends both on the strength of gravity, which holds the star together, and the strength of electromagnetic forces, which controls how efficiently energy is transported through the star and radiated into space. The details are complicated, but when they are worked out, it turns out that the lifetime of a typical star, in natural subatomic units, is precisely the ratio of the strengths of the two forces – 10^{40} – give or take a factor of ten or so. The conclusion is that *whatever* number this ratio might have been, intelligent creatures would only be around to wonder about it when the universe had existed for roughly this same number of subatomic time units.

We can go further than this and consider why this number is so big: that is, why gravity is so weak compared to electromagnetic forces. Our existence on Earth depends on the sun remaining stable for the several billion years that it has taken for biological evolution to produce intelligent creatures. Hence, the lifetime of a typical star like the sun ought to be at least this duration, which forbids gravity to be appreciably stronger than it is. If it were so, the sun would have burnt out before humans could have evolved.

The strength of gravity is also intimately related to another fundamental feature of our universe – its size. Most people are aware that the universe is big. First, the distances between stars are enormous. The *nearest* star to the sun is nearly thirty thousand billion miles distant (more than four light years), and the Milky Way galaxy is one hundred thousand light years across. Our telescopes can detect galaxies that are several *billion* light years away. Second, the sheer number of stars is mind-boggling. Our galaxy, a typical one, has about one hundred billion stars, and many billions of galaxies are known to exist.

There is a sense, however, in which the universe is limited in size. Crudely speaking there is an 'edge' about fifteen billion light years away. This is not a real physical edge, but is the horizon mentioned on page 154 beyond which the curvature of spacetime will not let us see. In this sense there is a natural size to the universe, and one can ask for a measure of this size in the smallest unit available – the size of an atomic nucleus. The answer is again about 10^{40}, but this time it is no surprise. We are really calculating the same quantity as the age of the universe in natural time units, only using distances (light years) instead of times (years). Thus, the universe is so extensive because it is so old, and it is so old because of the time taken for life to evolve.

Turning now to the contents of the universe, we can total up the quantity of matter, using the smallest unit of matter available – the atom. The number of atoms in the universe (inside our horizon) turns out to be about 10^{80}, or one followed by eighty zeros, which is just the square of the other large number (10^{40}) already discussed. It is possible to confront this further 'coincidence' also using the anthropic principle, because it happens that the total amount of matter in the universe is related to its age. The reason for this is that the universe is expanding and the density of matter controls the expansion motion. If the amount of matter were much greater, its gravity would halt the expansion and cause the universe to collapse before intelligent life could evolve. On the other hand, if matter were sparser the expansion would be more rapid. If this were so it would be unlikely that galaxies or stars would ever arise in abundance. As mentioned, galaxies and stars form out of concentrations of gas and dust whose local gravity pulls surrounding material together more strongly than the expanding universe can disperse it. If the density of matter in the universe were much lower, the local gravity would be less, and would be unable to restrain the receding matter. In addition, the expansion rate itself would be higher, making the competition still less favourable for the formation of dense regions. It seems that we could not exist in a universe that was very different in density from the one we actually inhabit.

If life is to exist, the density of the universe must be high enough to trap matter locally into stars, but not so high that the whole cosmos collapses. We can use Einstein's general theory of relativity to calculate the optimum density that strikes a compromise between the two alternatives, and use this density together with the size of the universe to calculate the corresponding total number of atoms. The calculation itself is elementary, and the answer can be expressed as the age of the universe divided by the gravity of one atom. This expression is in fact numerically very nearly the same as multiplying the two ratios discussed on page 172: the age of universe times the ratio of electric attraction to gravity in the atom $=10^{40}\times10^{40}=10^{80}$. This is just the observed number of atoms. Therefore, yet another 'amazing coincidence' turns out to be no surprise after all, given that we are alive to remark on it.

Similar arguments to that of gravity have been proposed in connection with the nuclear force. We saw in chapter 8 that the stability of nuclei depends on a balance between nuclear attraction and electric repulsion, so changes in the strength of either force would threaten the structure of composite nuclei on which life depends. For example, an increase of ten in the fundamental electric charge carried by protons would be enough to disintegrate carbon nuclei; a corresponding decrease in the nuclear force would have the same effect. Fred Hoyle has pointed out that the existence of carbon may be still more delicately dependent on nuclear forces, because according to the big bang theory, the present structure of the universe could not have withstood the enormous temperatures of the primeval phase. Even atoms and their nuclei would have been smashed by the heat energy, so in particular, as already remarked, carbon was absent. Before the first few minutes had elapsed, temperatures in excess of billions of degrees ensured that only individual protons, neutrons and other particles existed; no composite nuclei could have held together because of the intense heat. As the universe cooled off, composite nuclei began to form, mainly by the fusion of protons and neutrons into helium. Calculations show that about a quarter of the material ended up as helium, but almost no heavier elements formed. The reason for the incompleteness of the nuclear synthesis is because after a few minutes the temperature had fallen too *low* for nuclear burning to proceed. The universe had only a few minutes of time between the searing heat of the furnace and the plummeting temperatures of the cooling plasma during which it could cook complex nuclei. That was not long enough for much output, which explains why the universe is made almost entirely of hydrogen and helium.

Carbon, the vital life-giving element, was synthesized much later when primeval-type temperatures were re-established in the centres of stars. Carbon forms only after much of the star has already been converted to helium. A nucleus of carbon consists of six protons and six neutrons, while that of helium contains two protons and two neutrons, so carbon can form when three helium nuclei collide simultaneously. In the hot stellar interiors collisions occur in profusion as the particles rush around chaotically, but a triple encounter is obviously much more rare than a two-nuclei collision. The fusion of three helium nuclei into one carbon nucleus is therefore a slow-going affair and might not have got off the ground at all were it not for an apparently fortuitous fact. The three helium nuclei fuse in two stages: first two of them unite temporarily to form a nucleus of beryllium. This union is only short-lived, and the success of carbon synthesis depends on the further capture of a third helium nucleus being fairly efficient. The efficiency of nuclear capture can vary enormously with energy, being greatly enhanced if the composite body is left with an energy close to one of its natural, internal, quantum energy levels. Hoyle pointed out that beryllium plus helium does indeed possess an energy level very close to the average energy found at the centre of a hot star, and this seemingly coincidental fact is responsible for the abundant production of carbon, which is subsequently dispersed through space when the star explodes. Moreover, it is important that once the carbon has formed, it is not immediately destroyed again by further synthesis and helium capture. Fortunately, however, no energy level exists in the composite carbon-helium system (which is in fact oxygen), so the subsequent depletion of carbon to generate still heavier elements is rather low. The energies at which these vital levels occur depends on the strength of the nuclear forces, so even a slight change of a few per cent would spell disaster as far as carbon-based life is concerned. If the nuclear forces take all manner of values in the other worlds of superspace, it is clear that only those universes, like ours, in which very special values are found can support abundant carbon-based life.

Another delicate dependence of life on nuclear forces has been mentioned by Freeman Dyson. Among ordinary hydrogen is a small fraction of what is known as heavy hydrogen, or deuterium. This is chemically identical to ordinary hydrogen, but its nucleus contains not just a proton, but a proton and neutron combined. Theory indicates that there is a tight competition between the quantum zero-point energy mentioned on page 67, which tends to prevent the neutron from being trapped, and the nuclear attraction force. In the case of deuterium, the attraction just wins – and experiment confirms this is so – the deuterium nucleus is only just bound. If two protons came together, the story is different. Protons have to contend with electric repulsion, and also the effects of the Pauli exclusion principle (see chapter 4) which prevent two protons from coming too close. In the case of the di-proton, repulsion wins, and it just fails to form a bound union. However, if the nuclear force were a fraction stronger (even a few per cent) the di-proton would become a reality. It would not remain a di-proton for long, because there is an energy bonus if one of the protons converts into a neutron by the beta-decay process, thereby transmuting the di-proton into a nucleus of deuterium.

Dyson examines the effect of these possibilities on the nuclear processes that occurred in the primeval universe, and points out that all the material which is now in the form of hydrogen would have formed di-protons and thence deuterium just after the big bang. With deuterium rather than hydrogen as the raw material, the primeval furnace would have processed the nuclear fuel at a vastly enhanced rate, gobbling up all the deuterium into helium nuclei, leaving a universe of virtually one hundred per cent helium. Stars like the sun, that spend billions of years quietly burning hydrogen in a stable condition, would not exist. Nor would water (which is hydrogen dioxide), upon which life as we know it is crucially dependent. It seems that life depends critically on that near-failure of the di-proton.

The other nuclear force – the so-called weak interaction, responsible for beta radioactivity – is also vital for life in the universe, in two ways. The first of these concerns the constituents of the primeval material, from which helium was synthesized in the first few minutes. Helium consists of two protons and two neutrons, so the amount of helium which forms depends on the proportion of neutrons present in the early stages. In fact, nearly all the available neutrons present in the primeval material become incorporated into helium nuclei, so the hydrogen from which most of the universe is made is in fact just a residue of protons that remained unpaired by neutrons because of a paucity of the latter. The heat energy of the primeval furnace is shared out among all the species of subatomic particles, and in the very early stages a balance is struck between the amount of energy that is used to make protons, and the amount that is used to make neutrons. This equilibrium is maintained because of the weak force: if there is an over-abundance of neutrons, some will use the beta radioactivity process to convert to protons and vice versa, bringing their proportions back into balance. It is a servo-mechanism that operates efficiently so long as external disturbances do not disrupt it, but account must be taken of the fact that the primeval material is embedded in an explosively expanding universe. At first the expansion motion is unable to tilt the balance, because the neutrons and protons are so hot and densely packed. At around one second, however, the density and temperature have fallen low enough – a mere ten billion degrees – so that equilibrium cannot be maintained, and the neutron-proton abundance ratio remains frozen at the value it had at this time. Calculations show the ratio to be around fifteen per cent, which yields about thirty per cent helium to seventy per cent hydrogen, precisely the abundances observed today.

The reason why the strength of the weak force is important here is because it also controls the moment at which equilibrium starts to fail. If the force were weaker, it could not hold the balance for so long in the face of the rapid expansion. This is vital because at earlier moments than one second there was a greater proportion of neutrons present, due to the following considerations. Neutrons are about 0.1 per cent heavier than protons, so they use up more energy in the making. If the available energy is sparse, this mass difference favours protons over neutrons, which is why at one second there are eighty-five per cent protons to fifteen per cent neutrons. At earlier moments, however, the temperature is higher, so more energy is available to share between them. The competition of the masses is then not so fierce and they get about equal amounts, leading to roughly fifty per cent/fifty per cent, or half protons, half neutrons. If this were the abundance ratio when equilibrium failed, it would lead to a hundred per cent helium production, because each proton would be paired with a neutron and no residue of free protons would remain to form hydrogen. As already remarked, a universe devoid of hydrogen would have no water and no long-lived, stable stars, which is a grim outlook for life.

The second place where the weak force is vital for life concerns the demise of massive stars. Some stars, having synthesized elements such as carbon and oxygen in their interiors, start suffering from fuel starvation. The crisis comes on slowly but escalates, until the core of the star cannot generate enough heat to prevent itself collapsing under gravity. The result is a progressive shrinkage followed by a sudden and violent implosion which liberates titanic forces. In particular, huge quantities of neutrinos, which are subatomic particles so tenuous that they easily pass right through the Earth without noticing, are liberated and cascade out of the collapsing core. So high is the central density of the star – about a million billion times that of water – that even these ephemeral particles are impeded in their outward surge. The precise value of this resistance to the neutrino flood depends on the strength of the weak force, which controls the interaction of neutrinos with other matter. If it were much stronger these neutrinos would be unable to escape at all from the core.

When they reach the overlying layers, the neutrinos blast them out into space in a horrendous explosion which illuminates the whole galaxy, pouring billions of times as much energy into space as a normal star. This awesome event is known as a supernova, and among the shattered debris of the star are found the elements such as carbon and oxygen. These elements are eventually scooped up into other star systems, to become the raw material out of which planets, and life, will form. Thus our bodies are made from the ashes of long dead stars. If the weak force were much stronger, so that the collapsing cores retained most of the neutrinos, no supernova explosion would result. On the other hand, if the force were weaker, the neutrinos would lose their grip on the overlying layers and fail to blow them into space. Either way spells a world starved of raw materials and, presumably, life.

There are probably many more features of the world that are vital to the existence of life and which contribute to the general impression of the improbability of the observed world. We have no idea, for example, why there are only three dimensions of space and one of time. Mathematical physicists frequently examine how the laws of physics would differ if the dimensionality were different, and there is no doubt that the world would be a very strange place if there were only two space dimensions, for instance. Whether life would be impossible is not known.

We do not understand why the subatomic particles have the masses that they do rather than some other values. Certainly we know that if the electron mass were, say, ten thousand times smaller, then the atomic orbits would start to collide with the nuclei and chemistry would be drastically altered, but why it cannot be slightly different is a mystery. Perhaps the values are just random and have no significance, or perhaps one day they will emerge as consequences of some fundamental theory, and so be forced to take the values they do.

Mankind's perspective of his place in the universe is bound to be influenced by the answer to the question: how special is the universe? In previous centuries when religion provided the foundation for man's conception of nature, it was taken for granted that it was very special indeed. As remarked in chapter 1, the earliest cultures understood few real laws of nature anyway; almost all phenomena were attributed to specially motivated gods and spirits, so even the routine affairs of the world were engineered around mankind. With the Newtonian revolution the opposite position gained ascendancy; the world as a machine which ticked over lawfully from aeon to aeon, completely predetermined by conditions in the infinite past, and uninfluenced by the cares and aspirations of man. Modern cosmology, however, postulates a creation at a definite time in the past, and the question re-emerges as to whether this event is in some sense a random accident or a highly organized spectacle.

Throughout history people have fallen into the trap of attributing special organization to the world where none existed. The gods of our forebears manipulated the world and maintained its activity. Modern science disposed of the gods and replaced them with laws of nature. Darwin even removed divine influence from the realm of biology. In the twentieth century most of what was formerly regarded as miraculous is now seen as an inevitable consequence of natural laws. The existence of the Earth is no longer regarded as extraordinary, for we understand, at least in outline, the mechanisms which brought the Earth into being; we also know when it happened. Even the existence of the sun is no miracle, for we can observe stars being born right now by turning telescopes on distant nebulae. Man, once regarded as the greatest of all miracles, is seen as a point on the road of evolution that began three and a half billion years ago and will, all being well, continue for a few billion years yet. Astronomers envisage planets throughout the universe on which alien life forms have arisen naturally as a consequence of the laws of physics and chemistry; probably there are many life forms of far greater intelligence than our own with technological communities incomparably more advanced than on Earth.

In summary, science has answered many fundamental questions

about how the world got the way it is, so that at least in outline we may write a history of the first fifteen billion years of the universe, starting at one jiffy. The central message is that nothing miraculous or remarkable has occurred on the way, other than the incomprehensibly remarkable fact that there is anything at all. We do not understand why the laws of physics are what they are, though we can marvel at their awesome beauty and mathematical simplicity. But given those laws, the world we perceive seems to follow naturally and automatically out of the big bang.

It is into this tidy scheme that the considerations of the last two chapters intrude, because although there is nothing remarkable about our local region of the universe – life on Earth, the solar system, even the galaxy – when it comes to the global features we find some very surprising circumstances indeed. The gravitational arrangement of matter in the big bang was apparently organized with a precision that surpasses belief. Whereas previous generations have marvelled at the delicate organization of our planet, this generation can take the planet for granted but marvel at cosmology instead. Why this organization appeared during the big bang we have no idea.

Different people will interpret these results in different ways. For those whom the religious explanation of nature still plays a part, the primeval cosmic order will be regarded as a manifestation of God's purpose, fashioning the universe for habitation in a very special way, much as the biblical writers interpreted it on a more parochial scale. Some scientists will be confirmed in their belief that this is not the only universe, but one of countless billions in most of which less remarkable things occur. These other worlds need not be elsewhere in superspace. They could, for example, exist in regions of space so distant that we cannot see them, or they may occur in the remote past, or in the far future, when the present scheme of things is finished.

John Wheeler, who invented superspace, envisages a universe which continues in its expansion until some final moment, after which recontraction occurs, sweeping all the galaxies back upon one another until they obliterate in a gigantic cosmic cataclysm like the big bang in reverse. In the extraordinary world of Jiffyland into which the cosmos returns, all of our physics could be reprocessed, so that if the universe could somehow avoid a singularity and emerge again, it would be with a new set of numbers, a different degree of primeval turbulence, perhaps new values for the strength of gravity and the other forces – even new laws of physics. So it would continue in cycle after cycle of activity – expansion and recontraction – with a sort of 'new deal' universe every time. Most new deals would be a poor deal for life, though, as the universe is endowed with unpleasant features according to the laws of probability. Eventually, however, against all the astronomical odds, the numbers would come out right purely by random shuffling, and that particular cycle would become inhabited, spawning intelligent creatures and cosmologists. If we believe that there are countless other universes, either in space or time, or in superspace, there is no longer anything astounding about the enormous degree of cosmic organization that we observe. We have selected it by our very existence. The world is just an accident that was bound to happen sooner or later.

Finally, there will be those people who will not entertain the idea that other universes really exist. They will either concede that the world has a stupendously fortunate structure as far as we are concerned, but accept this as a fact of nature much as it is accepted that the sky is blue, or they will challenge the whole philosophy and seek to demonstrate that there is, after all, nothing very special about the organization of the primeval universe. To establish this counterproposal it will be necessary to show how the high degree of uniformity in the arrangement of matter and the cosmological motion has arisen automatically from certain physical processes in a way that avoids generating vast quantities of heat. What this amounts to is standing on the assertion that 'all things are not equal' as discussed on page 168. If new physics comes into play to prevent the more likely black hole states being tried, and to direct the activity of the universe to star-making instead, then it will no longer be surprising that the universe is not dominated by black holes. Similarly, if certain as yet unknown physical mechanisms prevent the turbulent motions from romping away and divert all the explosive expansion energy into the smooth and orderly dilatory motion that we observe, then the vast quantities of heat that would have otherwise accompanied the turbulence will not appear.

At present it is not possible to give a definite answer to these questions, because so little is known about the physics of the very early universe; the extreme conditions encountered there are way beyond all present experimentation and most mathematical calculation. But if one cannot claim unequivocally that the universe is organized into shape rather than driven there automatically by new physics, one can at least direct attention to the issues at stake. For centuries mankind has grappled with questions of existence: his own and the relationship between himself and the existence of the universe. With our scientific knowledge we can see this problem in a new light. Man is not just a spectator in the universe, an incidental feature carried along on the tide of events of the cosmic drama, but an integral component. Whether or not new knowledge of the primeval cosmos will change our conclusions about how it all began, we know at least that we are playing our part.

10. *Supertime*

And, departing, leave behind us
Footprints on the sands of time.

H. W. Longfellow 1807–1882

In the earlier chapters a great deal of space has been devoted to the role of man as observer in the universe. In particular, the nature of reality and perhaps the very structure of the universe is intimately related to our existence as conscious individuals perceiving the world around us. Accepting this central role for man in nature runs counter to all previous scientific development which has demoted him from the pinnacle of the creation to a run-of-the-mill bioform. Yet deep mysteries remain about the mechanism of perception and the nature of consciousness as such. Is awareness of the environment and one's own existence a feature of human life only? Is it restricted to the primates? Is it possessed by all animals, all life?

Dealing with matters of consciousness and perception is alien to the whole tradition of physical science, which usually attempts to abstract away from the individual observer and treat only objective reality. Repeatable experiments, measurements conducted by and recorded on machines, mathematical analysis of the results and other techniques have been developed to exclude the experimenter himself as much as possible from science. However, we have seen in the foregoing chapers that 'objective reality' is an illusion and that the all important laboratories and machines owe their very existence to the human experimenter whose existence in turn may be interwoven with the fundamental features of nature and the organization of the cosmos. Sooner or later the observers – us – intrude into the picture.

If we take consciousness seriously we are faced with the conundrum that nobody has succeeded in registering its existence in an experiment. That is to say, the human brain has been much explored and a great deal of its workings understood, but so far it has not been experimentally demonstrated that consciousness is needed as an additional component in the operation of the brain. Some scientists believe that consciousness *is* the operation of the brain, and that is all that need be said. To others, this idea seems manifestly absurd. We saw in chapter 7 how one scientist at least actually invokes consciousness as a definite physical system, over and above the brain, as a mechanism for collapsing the quantum state into reality.

Whether or not mind exists as distinct from brain processes, there are mysteries about the very nature of our elementary perceptions. Never is this more true than in our perception of time. The theory of relativity was outlined in chapter 2 where it was explained that physicists think of the world as four dimensional: three of space, one of time. Lines thread themselves through this spacetime continuum, representing the histories of bodies as they perform their processes. The lines are not independent, but interact through a variety of forces. We see a gigantic network of influence and response filling the universe and stretching from past to future. That *is* the universe.

This is not at all the image of time as we perceive it. Looking out on the world about us we see the whole drama being acted out, as event after event unfolds. Our view of the world is a movie: things happen, changes occur, future events come into existence and pass away again. In short, it seems to us that time passes. How can this kinetic image of the world that we actually experience be reconciled with the static picture of spacetime simply being there?

Let us analyse more closely the nature of time as we perceive it. In daily discourse we entertain two rather distinct, and possibly incompatible, images of time which nevertheless coexist in our minds without causing many people any intellectual difficulty. First we label events by dates: the battle of Hastings (1066), the election of President Carter (1976), a total eclipse of the sun in Britain (1999), the chiming of my clock (3 pm, 12 November, 1980). Time is like a line stretching off into the dimness of the past and into the distant future, each point on the line carrying a date which is itself just a label giving the duration in, say, years from some arbitrary event, such as the birth of Christ, deemed to be of special significance to the community. Relabelling the dates, for example by adopting the Jewish or Chinese calendars, doesn't change the events or the relationship between them, and is as harmless as using metres rather than feet to measure distances.

Associating events with dates is entirely equivalent to associating places with map references. In this respect the date-label view of time is the one adopted by physicists, where time is simply there, stretched out like a line, full of interesting events from the moment of the big bang to future infinity (or the big crunch, if there is one). There is, to be sure, a subtlety of which physicists are aware that is ignored in daily discourse, and that is the fact that time is relative to the state of motion of the observer. In chapter 2 we discovered how the notion of simultaneity – two events having precisely the same date – is meaningless unless they are located at the same place. Observers moving differently will disagree about whether two events are simultaneous, or in sequence, so they will assign them different dates. This complication is not a fundamental problem so long as we know the rule which connects one observer's set of dates with another, so that we can interconvert their observations. The rule is indeed known, and supplied by the mathematical formulae of Einstein's theory of relativity. Moreover, the rule works spectacularly well, as many laboratory experiments on time have demonstrated.

Quite apart from our labelling events with dates we use an entirely different mode of language and system of thought about time which is based on the kinetic image: the system of tenses. We say (and think) that the battle of Hastings *occurred* in 1066, that the eclipse of the sun *will happen* in 1999, and that my clock is chiming *now*. The past, present and future are so fundamental to our perception of time that we normally accept them without question. Because of this view, time acquires a much richer structure than mere date labels can assign to it. Firstly, it is divided into three sets. The future, which is unknowable and perhaps in part amenable to our will: it contains events which do not yet exist and which may not even be determined due to quantum uncertainty, but which eventually will come into being. The past, which can be known and is partly remembered, contains events which have occurred which we are powerless to change, however strongly we wish to. The events once existed but have passed out of existence into a sort of fossilized inaccessibility. Finally, where past and future join we have the present – *now* – a mysterious and fleeting thing, without noticeable duration, which bestows upon the events that are simultaneous with it a kind of concrete reality that is not possessed by the ghostly and insubstantial images of past and future events. The present is our moment of access to the world – when we can exercise our free will and alter the future. This image of a special status ascribed to the present is echoed in Longfellow's words: 'Act, act in the living present!' Our vision of reality is, therefore, strongly rooted in the tense structure of time.

Division of time into *the* past, *the* present and *the* future is a far more elaborate organization of ideas than the simple relationships between dates, such as the assertion that Carter was elected *after* the battle of Hastings or that my clock chimes *before* the eclipse of the sun. These latter couplings express before-after relations that are quite independent of the moment in time when they are examined. Carter after Hastings was always true, is true now, and will always be true in the future.

So far, it may appear that there is nothing especially incompatible about the co-existence of dates and tenses. The paradoxes creep in, however, when it is appreciated that the system of tenses is not static, but moving. The present, which we usually identify as our moment of conscious awareness, is steadily moving towards the future, coming across new events and consigning the previous ones to memory and history. Alternatively we can view the 'now' of our awareness as fixed, with time flowing past our consciousness in a stream, sweeping away the past and rushing the future towards us. Either way, the impression of a flowing, moving, passing time imbues the world of our experience with change and activity.

What is the passage of time? In literature, art and religion it has been expressed many ways. Most frequent is the analogy to a river; Saint Augustine (354–430) expressed it thus:

'Time is like a river made up of events which happen, and its current is strong; no sooner does anything appear than it is swept away.' For H. D. Thoreau (1817–62) 'Time is but the stream I go a-fishing in'. Sometimes the image of flight seems closest, for Virgil 'Time is flying – flying never to return', while Andrew Marvell (1621–78) sees time as a 'Winged chariot'. Robert Herrick (1591–1674) advises us to 'Gather . . . rosebuds while ye may, Old time is still a-flying'.

William Shakespeare returns often to the theme of time's passage. In *Twelfth Night* it is a 'whirligig' which 'brings in his revenges', and this destructive or vengeful element is a favourite. Byron speaks of 'Time the avenger'. Ovid describes 'Time the devourer of things', and Tennyson warns that 'Time driveth onward fast . . . All things are taken from us and become, Portions and parcels of the dreadful Past'. Herbert Spencer (1820–1903) defines time cynically as 'that which man is always trying to kill, but which ends in killing him'. All these images elaborate our deepest impression of time as irreversible motion implementing change.

When it comes to science, the images are not so graphic. Scientists, like everyone else, use tenses both in daily life and in discussing experiments and observations about the world, and yet in their theoretical analyses of nature, tense plays no part: there are only dates. Nothing appears in Newton's equations which corresponds to a present, no quantity appears that gauges the motion of time. Of course, time itself is there, and the equations predict what events (e.g. dropped apple reaches the ground) occur at what time, but we cannot use Newton's equations, or any others in science, to tell us *what time it is*. In experiment as much as in theory, the laboratory is powerless to reveal the flow of time, as no instrument exists which can show its passage. As remarked in chapter 2, it is wrong to assume that a clock is such a device. A clock is only a method of assigning dates to events, though we perceive the operation of the clock as motion, but motion in space (i.e. round the clock face) rather than in time. It is our psychological impression of a moving time that, because of the close association of clocks with time, falsely bestows upon the clock the appearance of measuring the passage of time.

The nebulousness of the moving time concept is well exposed by asking how fast time flows. What mechanism do we possess to measure the speed of time? If such a machine existed one could consult it each day to find out if time had gone more slowly that day, or if the pace of events was quickening. Most people's perception of time has this variable character. It is a common experience that ten minutes in a dentist's chair seems like half an hour of some more pleasurable pastime or that a day full of activity passes more rapidly than one spent in idleness or boredom. These are, of course, psychological impressions related to one's personal mental state. The rate of passage of time will always be one day per day, one hour per hour, or one second per second. Even boring days take one day to pass. It is meaningless to say 'today took only tweve hours' for what is really meant is 'today *seemed* like only twelve hours'.

If one insists on maintaining the notion of a moving time then there appears to be a blatant incompatibility between tenses and dates. Dates are fixed to events once and for all, but the tense labels of the events change from moment to moment. Thus, Carter's election was a future event in 1975 and is a past event now.. How can the same event, with a fixed date, be past, present *and* future? Evidently 'pastness', 'presentness' or 'futurity' are not qualities that are intrinsically possessed by events, nor can they even be made very precise, for if we ask when an event has pastness and reply 'When it has happened', that is a mere tautology. How do we know it has happened? Because it is in the past. The analysis goes round in circles.

Presentness is equally intangible, for what *is* the present? We are certainly agreed that the present is a single moment (or at least a duration so short we cannot perceive any internal structure), but which moment? The answer is, of course, every moment. All moments are the present moment, when they happen. But when do they happen? At that moment! The discussion gets nowhere. Even after profound introspection one concludes that nothing of any substance is being said, that the qualities of past, present and future are so manifestly obvious, so fundamental to our experience, that we cannot begin to approach their immediacy by putting the experience into words. Saint Augustine expressed this dilemma when he said that he knew what time was, so long as nobody asked him to explain. Then he did not know. Charles Lamb (1775–1834) expressed the feeling thus: 'Nothing puzzles me more than time and space; and yet nothing troubles me less, as I never think about them.'

Feeling that time really does pass, and that there exists a past, present and future does not contribute one bit to our understanding of the objective world, yet these concepts are indispensible to the organization of our personal affairs and the conduct of our daily lives. Are they just complete illusions, or do our perceptions probe a structure of time – a supertime – that is as yet unrevealed in the laboratory? Does true reality depend upon the existence of a present moment?

These questions provide one of the great challenges to modern science and philosophy, and there is no measure of agreement, even about how to formulate the concepts. Nevertheless as the earlier chapters of this book have shown, recent developments in quantum theory and cosmology have begun to touch upon these topics, and we are approaching the time when they will have to be squarely faced.

Let us examine in turn the two opposing viewpoints, taking first the objectivist stance that would probably be taken by most scientists and many philosophers. According to this point of view time does not pass, and the past, present and future are merely linguistic conveniences, with no physical content. In spite of its startling implications, this position can easily be defended. The central argument is that there are dates, and events associated with them. The events have past-future relations, but they do not *occur*. In the words of the physicist Hermann Weyl 'The world does not happen, it simply *is*'. In this picture things do not change: the future does not come into being and the past is not lost, for all of past and future exist with equal status. Shortly we shall examine how quantum theory squares up to this apparently deterministic picture but for now we can note that if one subscribes to the many-universes interpretation of quantum theory then there exists not one future, but trillions of them, namely, all the subsequent branches from this moment. In spite of this complication, the basic argument is unaffected.

The surprising thing is that the above picture appears so alien and outrageous, because it is so obviously correct in its individual statements. The sceptic would counter that of course things happen; there *is* change. 'Today I smashed a teapot: this event occurred at four o'clock and it is a change for the worse. My teapot is now broken.' But let us analyze what the sceptic is really saying. Before four o'clock the teapot is intact, after four it is broken; at four it is in a transitional state. This mode of language – the physicists' date-label language – conveys precisely the same information, but in a less personalized tone. No compelling need is found for the mode of description that the intact teapot *changed into* a broken one at four o'clock, or that the event *happened* at four. There are dates, and states of teapot. No more need be said.

'Ah!' retorts the sceptic, 'perhaps I don't need to use moving-time language, but that is the way I perceive the world, that is my psychological impression of time: I feel it passing.' This is a legitimate and obviously correct comment because we all share the fundamental feeling of things happening around us, and time passing by. Yet it is dangerous to base too much of our science on psychological feelings, for we know of many instances where they lead us astray. We all feel the moon is larger when near the horizon than high in the sky, but it is not; we all feel a vertical drop of one hundred feet is longer than the same distance horizontally; we all feel that the Earth is fixed, but it is in reality moving; and so on. Should we have any more confidence in our feelings about time than we have in our feelings about spatial arrangements and motion?

Internal feelings of flux and motion are easy to generate. By spinning around a few times, the fluid in the organ located in the inner ear which helps the brain to maintain a sense of orientation and balance can be set in motion. On stopping, the impression of rotation persists forcibly: we feel dizzy. One can look fixedly at a point on the wall and convince oneself rationally that the world is not rotating or rushing past. Yet however convincingly it can be seen that the wall remains still, the motion is *felt* among one's perceptions. One might wonder why the motion is, say, anticlockwise rather than clockwise, in direct analogy to the problem of why time always flows from past to future. There seems to be no strong reason for supposing that the flow of time is any more than an illusion produced by brain processes similar to the perception of rotation during dizziness.

Accepting the passage of time as an illusion makes it no less important. Our illusions, like our dreams, are very much a part of life. They may have no 'objective reality' but we have come to see that such a thing is in any case a rather vague concept. According to the physicists' static picture of time we need not be remorseful about the past or concerned about the future. Death, for example, need hold no more fear than the state 'before birth'. If there is no change, people cannot die in the strict sense of the word. There are only dates when an individual is alive and conscious, and other dates (before birth, after death) when he is not, and nobody can be conscious of unconsciousness, for that is a contradiction in terms. It might be objected that we are only conscious of one particular moment, and that moment marches inexorably forward, so that when it reaches death, all is lost and experience ceases. However, it is not true that we are only conscious of one moment, for we are obviously conscious of every moment of which we are conscious! To counter that we are only conscious of one moment 'at a time' is an empty remark, because obviously each moment is separate from all the other moments. Our experience cannot move along our lives, for each moment of our lives is experienced. Every moment of our lives is regarded as 'now' by the mental state that we associate with it. There can be no single 'now' and no individual 'present' because all experienced moments are 'nows', and all experiences have 'presentness'.

In spite of the manifest truth of all these remarks, one is still left with a profound feeling of dissatisfaction that something is missing. Indeed, the desire to find that extra ingredient, something upon which to build the flow of time and the existence of the now, has plagued physicists for years. Some have turned to cosmology for the answer, others to quantum theory. At first the indeterminacy of quantum theory seems to offer a possibility, for if the future is still in the balance of chance there might be a sense in which it is less real than the present or past. Some physicists have compared the impression of the future coming into being with the collapse of the quantum superposition into reality. Superficially this appears promising, because the collapse process is known to be asymmetric in time (i.e. irreversible) in a very fundamental way, so it shares some of the features of memory. According to this belief, the present is a real phenomenon and represents the moment when the world changes from potential to actual, e.g. Schrödinger's cat is found to be either alive or dead, the moment of decision defining some sort of present. These ideas have also been used to argue for the existence of free will, a subject closely interwoven with our picture of reality and the nature of time. If the future is undetermined, perhaps our minds can act on the world at the quantum level and tip the balance of chance in the direction of our choice?

The argument might run something like this: The brain operates by arranging electrical impulses, and the electric currents consist of electrons moving in compliance with the laws of quantum theory, which means they will not be completely 'well behaved' but subject to random fluctuations and indeterminacy. Suppose that there exists, in addition to the brain, a mind which can act at the quantum level to determine which of the many possible paths certain crucial electrons end up following. The laws of quantum theory are not violated, for many paths are possible, the mind simply ensures that the one of its choice is realized. In this way, mind is capable of organizing brain states entirely within the laws of physics. The brain states in turn operate the body, which then manipulates the environment, so enabling the mind to gain control over the material world. Some researchers have even claimed to have measured the effect of mind on quantum processes by getting a subject to 'will' certain radioactive decays in an ESP-type experiment.

These ideas do not really stand up to scrutiny. The fact that the future is indeterminate does not necessarily mean that it does not exist, only that it does not follow slavishly from the present. In addition, the fact that we regard the future as indeterminate but the past as concrete is closely connected with the way we actually perform experiments and organize the results. Laboratory experiments involve preparation and analysis as well as actual experimentation, and this framework already imposes a past-future asymmetry on the interpretation of the results. Indeed, it is possible to perform a set of, crudely speaking, inverted experiments, in which instead of preparing a quantum state at the outset and measuring the result, the opposite is done: a number of results are collected and their initial states inferred. By reflecting in time the whole framework, asking different questions and analyzing different results, the past can be made indeterminate rather than the future. (In this scheme, the Everett branches fan out in the past rather than the future, so the worlds fuse together instead of splitting.) It follows that the different status of the past and future in quantum indeterminacy are not intrinsic, but reflect our attitudes to what is relevant, and the intellectual superstructure to which the experimental results are pegged, which in turn is a function of the strongly time-asymmetric nature of the world due to the thermodynamic processes going on around us. So once again, the impression of the future 'coming into being' seems to be an illusion based upon the lopsidedness of the world in time, and is not a real effect due to any motion of, or through, time.

Although quantum indeterminacy does not seem to offer a basis for an explanation of an objective flow of time, or the division of time into past, present and future, it could conceivably provide an explanation for the subjective experience of time if Wigner's interpretation of quantum theory is entertained. It will be recalled from chapter 7 that Wigner proposes to invoke the mind as the agent for collapsing the wavelike quantum superposition into a definite reality. It could then be argued that the mental impression of the passage of time passing is caused by the continual quantum collapsing going on in the mind.

As to whether the mind can in turn react back on the quantum brain to tip the balance of chance, there is no evidence (apart from the ESP experiments) that this occurs, and it would have to be demonstrated that the minute quantum effects involved can be amplified enough to develop signals at the sort of electrical level that the brain can use. Even if this were so, it is not clear that it amounts to genuine free will, or even that free will is meaningful, for if the mind itself is regarded as non-quantum and deterministic, and it decides to manipulate the brain to carry out a certain activity, then justification must be found for why the mind has embarked on that particular course of action. Because the mental state that initiates the action is completely determined by the mind's past states together with the influences impinging on it from the brain, then the mind will be reduced to a mere Newtonian automaton, with no control over its own affairs, its actions being entirely a consequence of past and present events. On the other hand, if the mind is indeterminate, after the fashion of quantum systems, then it will be subject to random fluctuations (uncontrolled whims) and arbitrariness will intrude into its decisions. Neither alternative seems close to the traditional concept of free will. The only truly free will would be if the mind could alter its own past states, thereby changing the present as well as the future. It is then free to construct whatever universe it wants, including itself, then to demolish it and rebuild it again, ad infinitum. Of course, in the Everett many-universes theory this happens in a sense, but there the freedom of the will is completely illusory, because all possible worlds actually occur and the mind splits repeatedly to populate vast numbers of them, each mind imagining that it is in control of its own destiny, but with all destinies really being achieved in parallel.

Although there is no hard evidence that the mind or will of the observer can influence the material universe by 'loading the dice' in the quantum game of chance, there is a sense in which the experimenter can decide the future. In chapter 6 it was explained how, by choosing from a number of incompatible observable quantities which one to measure, the experimenter can change the quantum alternatives on offer, even if he cannot force the choice. The example which was examined in some detail was the case of the polarizer and the photon, which enables the experimenter to create a world in which a photon has a certain particular polarization, should it make it through the polarizer. Another example concerns the position and momentum of a subatomic particle. By choosing which quantity to measure the experimenter can create a world where either the position or the momentum of the particle has a well-defined value, even though which value will be beyond his control and left as a matter of chance. It is rather like a lucky dip in which one can pick from either of two bins, the first containing a variety of chocolates, the second a variety of toffees. There is some element of chance and some of choice. It is important to realize that the power of the quantum experimenter to decide the future, limited though it is, is a great improvement on his pre-quantum counterpart who was purely an automaton, turning through time like the cogs of a machine. However, this facility notwithstanding, there is no reason to suppose that the future does not already exist, even though it is not yet determined, and even though the observer will have a hand in structuring it.

A final nail in the coffin of the idea that the future is waiting to come into existence is provided by the theory of relativity. As already explained, the simultaneity of spatially separated events is a relative concept, so it is clearly meaningless to claim that only the present is real, for whose 'present' is one referring to? The belief that the world out there only exists 'now' and that at the next moment it has 'changed into' a new condition and a new reality is badly misconceived, for not only is there no real world 'out there' anyway, as the analysis of the quantum measurement process demonstrates, but two observers who are in relative motion will ascribe completely different dates to the same events. For example, two people who stroll past each other on Earth will disagree drastically about which event on the distant quasar 3C273 happens to be simultaneous with their encounter. The discrepancy amounts to thousands of years. Each may assert the reality, at that moment, of their particular quasar event, but clearly this definition of reality is a useless one, being adjustable at will: it is merely necessary to get up from your seat and walk about to sweep through thousands of years of 'reality' on 3C273. A 'present' event there may be projected suddenly into the future, or past, and then be brought back again, solely by the expedient of ambling about. Similarly, sedentary aliens will be disagreeing with their pacing colleagues about whether it is 'really' 1980 on Earth, or the year 5760. Each will think that their particular choice of event is happening 'now' and is therefore real, while the other is mistaken. Neither is correct, for there is no universal present and no universal reality.

It would be exciting to identify the particular brain processes which are responsible for the feeling of temporal flux: it seems likely that they are closely associated with the memory process, which is also strongly asymmetric in time. We remember the past, not the future, so time is endowed with a sort of mental lopsidedness, and if we had no memory, consciousness would disappear along with the flow of time. I am not referring here to the condition of amnesia which only affects long term memory, but a state in which nothing at all, however recent, is remembered. In that condition one should be quite unable to make any sense of one's surroundings, for sensory information would be reduced to a collection of meaningless and incoherent momentary impressions, and planned action would become impossible, for one should be unable to recall from one moment to the next what one was doing or what the world was all about. Memory, at least on the short term, is an indispensible part of the perceptive process, as perception consists in organizing sensory impressions around previous knowledge and experience so that events can be related to one another and our own existence related to the world about us.

It might be objected that explaining the flow of time in terms of memory only replaces one mystery by another, for we have to account for the fact that only the past is remembered, and not the future. What is the origin of this past-future asymmetry? Fortunately we are on firmer ground here, because past-future relations are tenseless, and can therefore be examined within the framework of the known laws of physics. All around us are processes that display a strong past-future asymmetry. One of these has already been mentioned, i.e. the inexorable disintegration of order. The second law of thermodynamics states that the total chaos of the universe keeps on going up, so that the accumulation of order in one place must be paid for by a more than compensatory amount of disorder elsewhere. Thus, the accumulation of information in our memories is achieved at the expense of a great deal of bodily metabolism – the operation of sensory organs, the transfer and processing of the incoming data, the location of the appropriate brain storage facilities and finally the electrochemical rearrangement of the brain cells to register the newly-acquired facts. All these operations must be driven by the body, by utilizing energy released from food, which is an irreversible dissipation of organized energy into bodily heat, in conformity with the general principle mentioned on page 158. In conclusion, memory is not an especially mysterious phenomenon, and is possessed by systems other than humans, e.g. spiders and computers. Libraries and other inanimate records of the past, such as fossils, are all examples of memory in a general sense. All comply with the fundamentally time-asymmetric second law of thermodynamics, so all endow the world with a past-future lopsidedness that in our minds seems to get promoted to the more elaborate structure of a time which *flows* from past to future.

There are, of course, many other apparently irreversible phenomena around us that contribute to the world's lopsidedness or asymmetry in time. To take a few random examples: people grow old, buildings fall down, mountains erode, stars burn out, the universe expands, eggs break, ripples of water spread outwards from centres of disturbance, radio waves arrive after they are sent, perfume evaporates from open bottles, clocks run down. In all cases we never encounter the reverse sequence of events, such as clocks winding themselves up or radio messages arriving before they are sent. It is important to stress that these phenomena do not define *the* past or *the* future, which I have argued do not have meaning, but they do indicate which events are before or after other events. So, for example, if we take a movie film of an egg falling to the ground and breaking we have no doubt which end of the film represents the prior event, for in the real world eggs do not spontaneously reconstitute themselves: egg-smashing is irreversible.

A careful study reveals that most of the irreversible processes around us can be described by the general law of increasing disorder, i.e. the so-called second law of thermodynamics, that has been mentioned several times. In some cases, such as eggs breaking, perfume evaporating, mountains eroding or houses falling down, the increase in disorder is obvious. In other cases it is more subtle. A clock which runs down contributes to the general disorderliness of the world because its organized activity – the coordinated turning of the cogs and hands – disintegrates into disorganized activity, as the energy stored in the driving mechanism gradually dissipates as heat in the clockwork material. The energy originally stored in the spring ends up among random atomic jiggles rather than in the cooperative motion of the cogs.

It has long been a mystery as to why our world is asymmetric in time. Why does order always give way to disorder? To understand this very general tendency it is helpful to return to the example of shuffling a pack of cards. If the cards are initially in suit order and the shuffling is random, they will, with overwhelming probability, end up after shuffling in a highly disordered state. The odds against the shuffler just happening to reconstitute the suit order at the end are not zero, but incredibly small.

In many natural processes a kind of shuffling process takes place as a result of internal molecular collisions, as explained in the previous chapter. A good analogy to the pack of cards is provided by the example of the open bottle of perfume. At the outset the perfume, like the cards, is in a very ordered condition, i.e. confined to the bottle. Due to a hail of impacts from the surrounding molecules of air, the perfume gradually evaporates, as its own molecules are ejected from the liquid surface and percolate around the room, driven by the ceaseless bombardment of the air molecules. Eventually the shuffling is complete and the perfume is spread irretrievably throughout the air, its molecules chaotically mixed among those of the air. The effect of the shuffling has therefore been to convert the original ordered condition of the perfume into a highly disordered state, apparently irreversibly.

The tendency for order to change irreversibly to disorder presents us with a paradox because it is known that the individual molecular collisions themselves are all reversible, so no fundamental laws of physics would be violated if the perfume spontaneously migrated back into the bottle; yet we would regard such an occurrence as a miracle. If, when two molecules collide and bounce away from each other we could, by some device, intercept them and turn them back along exactly the same paths, they would rebound again back to their original positions. If this were done simultaneously for all the molecules of perfume and air, the whole system would run backwards, like a movie played in reverse, until the perfume was all deposited in the bottle. The possibility of this miraculous turn of events is also evident from the card shuffling exercise, for if we continue to shuffle without ceasing, sooner or later we would succeed in shuffling the pack back to suit order. The time required would be immense, but purely on the basis of the laws of probability, random shuffling must eventually produce every conceivable sequence of cards, including suit order. Similarly, molecular collisions will eventually produce an ordered state once more, with the perfume in the bottle, provided of course that the room is sealed off to prevent the perfume from permanently escaping.

The paradox is, why, if both forward and reverse order-disorder transitions are equally possible, do we always encounter perfume evaporating into rooms, mountains eroding, ice melting when heated, stars burning out, sandcastles washed away by the tide, etc., etc? To resolve the paradox we must ask in each case how the ordered state was achieved in the first place, e.g. how did the perfume get into the bottle originally? Not, we may suppose, by someone opening a bottle in a perfume filled room and waiting an immensity of time for it to congregate in the receptacle by chance; that would be a strategy as inefficient as the fisherman who opens a basket by the river and waits for the fish to jump in. In the real world, ordered states are selected out of our environment at the outset, they do not form at random. The world about us abounds with orderly structure, most of it due, in the case of the Earth, to the proximity of the sun which drives much of the organized activity on the terrestrial surface. The sun, and stars in general, are the supreme examples of organized matter and energy in the universe. As time goes on, the ordered energy locked up in their interiors gets dissipated away as the stars burn up their fuel and spread the energy round the cosmos as heat and light. The stars burn out, and the universe, like a gigantic clock, slowly runs down. Even on a cosmic scale order is remorselessly decaying into disorder in a billion variety of ways.

The asymmetry between past and future, rooted as it is in the oneway tendency for order to disintegrate into chaos, is seen to have a cosmological origin. To explain where the ultimate cosmic order came from, and hence account for this distinction between past and future, it is necessary to consider the creation of the universe – the big bang. The cosmic structure which emerged from the primeval furnace was highly ordered, and all the subsequent action of the universe has been to spend this order and dissipate it away. Plenty remains, but it cannot last for ever. The orderliness which drives the sun and stars, so vital for life in the universe, can be traced to the nuclear processes which ensured that the nascent cosmos was made mainly of the light elements such as hydrogen and helium, a feature caused by the rapidity of the primeval expansion which did not give the cosmic material long enough to cook heavier elements in the early stages. It also depends upon the relative smoothness of the cosmic material, so that prolific black hole formation immediately after the big bang should be avoided. So once again we discover how delicately life in the universe, and our existence as observers, depends on the right cosmic arrangement, namely, one that gives a sharp distinction between past and future based on primeval orderliness – an orderliness that reaches a pinnacle of complexity in living matter.

The intimate connection between our own existence, the asymmetry in time of the world about us and the initial cosmic order must be viewed in the context of superspace. We have seen that the orderly cosmos is one of only a very small fraction of worlds out of all those that are possible. Among the other universes are those in which disorder reigns throughout, and also those which start out in a disorderly state and then progress towards order. In such worlds time 'runs backwards' relative to our own world, but if they are inhabited by observers, one supposes that their brains are also subject to reverse operation, so that their perception of their universe differs little from our perception of ours (though they would regard it as contracting rather than expanding).

When the equations for the quantum development of superspace are examined, they are found to be reversible – they do not distinguish past from future. In superspace there is no distinct past and future. Some of the worlds certainly have a strong past-future property and these are precisely the ones that can support life. Others have a future-past, reversed, asymmetry and presumably are also inhabited. The vast majority, however, have no such peculiar distinction between past and future, so are quite unsuitable for life and go unnoticed. In the Everett theory, all these other worlds, including the reversed-time ones, really exist alongside us. In the more conventional theory they are possible worlds which, by incredible good fortune, did not come to exist, though they could still exist in the remote future or on the other side of the universe. It could be that our own cosy, highly ordered world is just a local bubble of equability in a predominantly chaotic cosmos, seen by us only because our very existence depends on the benign conditions here.

In this chapter the physicist's model of time has been contrasted with that of our personal experience, full as it is with weird psychological images and paradoxical motion. The grey area between mind and matter, philosophy and physics, psychology and the objective world is only on the threshold of exploration, yet any ultimate picture of reality cannot omit it. It could be that the images of time so dear to us – the existence of a present moment, the passage of time, free will and the non-existence of the future, the use of tenses in our language – will come to be regarded as only primitive superstitions that spring from an inadequate understanding of the physical world. Maybe our descendants will not make use of these concepts at all, in which case one imagines that they will arrange their lives very differently from our own. It is conceivable that advanced communities elsewhere in the universe have long since abandoned the notions that time passes or that things change, that there is a single present moving towards an uncertain future. We can only guess at the impact that this abandonment would have on their behaviour and thought, for without expectation, remorse, fear, anticipation, relief, impatience and all the other temporally-related emotions that we experience, their conception of the world might be incomprehensible to us. It is probable that if we encountered such beings we would be unable to communicate much of common understanding. On the other hand, it could be that for once our minds are more reliable than our laboratory instruments, and time really does have the richer structure that we perceive. In which case the nature of reality, of time, space, mind and matter, will suffer a revolution of unprecedented profundity. Either prospect is awesome.

Index

Index